Historians Will Not Look Back On Us Kindly

Julian Del Bel

Table of Contents

Foreword
- Acknowledgments
- Introduction: The Age of Discontent

Part I: Environmental Missteps
 Chapter 1: Ignoring the Inferno - Climate Change Denial
 Chapter 2: Oceans of Plastic - Marine Pollution Unabated
 Chapter 3: Choking Skies - The Neglect of Air Quality
 Chapter 4: Silent Extinctions - Indifference Towards Biodiversity

Part II: Societal Failures
  Chapter 5: The Wealth Chasm - Economic Inequality Explored
  Chapter 6: Connected Yet Isolated - The Technology Paradox
  Chapter 7: Mind Traps - The Mental Health Crisis and Social Media
  Chapter 8: Identity and Division - The Politics of Us vs. Them

Part III: Governance and Political Shortcomings
 Chapter 9: Democracy's Decline - The Rise of Authoritarianism
 Chapter 10: Post-Truth Politics - The Assault on Fact and Reason
 Chapter 11: Surveillance Over Society - The Erosion of Privacy
 Chapter 12: Precarious Futures - The Unraveling Social Contract

Part IV: Cultural and Ethical Reflections
 Chapter 13: The Culture of Excess - Consumerism Unchecked
 Chapter 14: Ethical Bankruptcy - The Erosion of Moral Values
 Chapter 15: Educational Erosion - Failing Systems and Lost Potentials
 Chapter 16: The Illusion of Progress - Questioning Technological Salvation

Epilogue: A Reflection on Our Era
- Afterword: The Historian's Burden

Foreword

In the hustle and bustle of today's world, a paradox of existential proportions quietly simmers beneath the surface of our daily lives. As our global village marches forward on the path of technological advancement and interconnectedness, there exists an underlying narrative of regression that escapes the mainstream discourse. This book, titled "Historians Will Not Look Back on Us Kindly," acts as a pivotal reflection of our times, holding a mirror to the contemporary human saga. It uncovers the myriad of crises looming over our civilization, crises that are often overshadowed by the dazzling allure of progress.

This work is not just a collection of observations; it is a clarion call for introspection and awareness. It challenges the readers to look beyond the veneer of technological achievements and societal advancements, urging them to confront the fractures and fissures that threaten the very fabric of our society. As you look into the chapters that follow, you embark on a journey that is both enlightening and unsettling, a journey that questions the conventional narratives of progress and prosperity.

"Historians Will Not Look Back on Us Kindly" serves as a catalyst for critical thought, encouraging a dialogue on the sustainability of our current trajectory. It compels us to reflect on the legacy we are creating and the world we will leave behind for future generations. In its essence, this book is a testament to the complexity of the human condition in the 21st century, encapsulating the dualities of progress and regress, of hope and despair.

As you navigate through this introspective odyssey, you are invited to peel away the layers of your understanding, to rethink what constitutes true progress. This foreword sets the stage for a profound exploration of our collective psyche, our aspirations, and the unintended consequences of our actions. In doing so, it beckons us to consider how history will remember us, and more importantly, how we can steer the course towards a legacy that humanity can be proud of.

Let this book be a mirror, one that reflects not just the world as it is, but also as it could be. May it inspire a renaissance of thought and action, leading us to forge a path that aligns with the deepest values of human dignity, sustainability, and collective well-being. The journey through "Historians Will Not Look Back on Us Kindly" is not just a literary voyage but a pivotal moment of awakening, a chance to redefine the narrative of progress for ourselves and the generations to come.

Introduction: The Age of Discontent

As we navigate the complexities of the 21st century, we find ourselves at a pivotal juncture in history. This era, perhaps more than any other, is defined by its stark contrasts: unprecedented technological advancements exist alongside profound societal setbacks; our world is more connected, yet divisions between us grow ever more profound; wealth and knowledge are at their peak, while inequality stretches to new bounds. The Earth, the foundation of our existence, faces relentless exploitation, casting shadows over the sustainability of life itself. It is against this backdrop that "The Age of Discontent" emerges—a period marked by its glaring contradictions, demanding our attention and action.

The subsequent chapters look into the varied failures of our era. This exploration is not driven by a spirit of cynicism but by a critical concern for the world we inhabit. We confront environmental missteps, highlighting not just our neglect of the natural world but also our collective denial of the consequences that such negligence portends. Societal failures are scrutinized, revealing the fraying of our communal ties and the diminishing of empathy and solidarity among us. Governance and political shortcomings are examined, shedding light on the erosion of democratic principles and the ascendancy of authoritarianism, alongside the pervasive manipulation of truth. Cultural and ethical considerations reflect on our dwindling moral guidance amid the whirlwind of consumerism and ethical relativism.

This introduction is an invitation to traverse the intricate landscape of our current global civilization, standing at the pinnacle of technological achievement while potentially plummeting to the depths of moral and ethical integrity. It calls upon us to acknowledge the discontents of our age—not to sink into despair but to ignite the spark of awareness, contemplation, and, ultimately, action. Although historians may not remember us with kindness, we retain the power to alter the course of our narrative through the decisions we enact today.

In our journey through this book, we aim not for indiscriminate blame but for a deep understanding of the roots of our collective challenges. By confronting these uncomfortable truths, we take the first step towards healing and change. This book stands as a testament to the conviction that within the realm of critique lies the potential for transformation—transformation that is crucial for the future well-being of our societies, future generations, and the planet at large.

Part I: Environmental Missteps

Chapter 1: Ignoring the Inferno - Climate Change Denial

At the threshold of the 21st century, humanity is confronted with an existential crisis unlike any other in its storied history: global climate change. This phenomenon, characterized by unprecedented shifts in global temperatures and weather patterns, has emerged as a formidable challenge to the survival of myriad species, including humans. Despite a robust and growing scientific consensus that underscores the severity and immediacy of this threat, and a cascade of warnings from environmental experts around the globe, a paradoxical and dangerous trend has taken root—climate change denial.

This chapter looks into the intricate nature of climate change denial, exploring its genesis, underlying motivations, and the far-reaching consequences of such a stance. Through a meticulous examination, we aim to unveil the complexity of this denial, its character, and the profound implications it holds for global policy, environmental conservation, and the future of our planet.

Roots of Denial

The genesis of climate change denial is deeply entwined with the interests of the fossil fuel industry, which, by the late 20th century, recognized the growing scientific unanimity on global warming as an existential threat to their operational paradigms. In an echo of strategies once deployed by the tobacco industry—infamous for its efforts to obfuscate the links between smoking and cancer—fossil fuel conglomerates embarked on a vast disinformation campaign. The objective was clear: to sow seeds of doubt about the veracity and gravity of climate change.

This disinformation crusade was meticulously orchestrated, aiming to muddle public understanding and debate about climate science. By financing pseudo-scientific research, manipulating media narratives, and leveraging political influence, these interests sought to create a veneer of uncertainty where little existed. The campaign was alarmingly successful, leading to widespread public confusion and significantly hampering the momentum for meaningful climate action.

As we explore the roots of climate change denial, it is crucial to recognize the calculated nature of this campaign, designed not out of ignorance but from a deliberate intent to protect economic interests. The ramifications of this denial are profound, stretching beyond mere policy paralysis. They have contributed to a delay in global action, exacerbating the environmental crises and casting a long shadow over future efforts to mitigate climate change impacts.

In the following sections, we will further dissect the mechanisms of denial, the roles played by various actors, and the impact of these actions on global environmental and political landscapes. The story of climate change denial is not just a cautionary tale but a clarion call for accountability, transparency, and, above all, urgent action in the face of a warming world.

The Machinery of Misinformation

At the heart of the climate change denial phenomenon lies a well-oiled machinery of misinformation—a network of think tanks, lobbying groups, and pseudo-scientific bodies designed to manufacture and propagate an alternative narrative on climate science. This network, sustained by substantial funding from industry interests, especially those tethered to fossil fuels, has been instrumental in producing and disseminating a deluge of content aimed at undermining the scientific consensus on climate change.

These organizations, often cloaked in the guise of academic rigor and intellectual freedom, have been prolific in their output. They churn out a continuous stream of reports, opinion pieces, and media appearances, each designed to cast a shadow of doubt over the realities of climate change. Their arguments are carefully crafted to appeal to a sense of historical continuity and naturalistic explanation, suggesting that the climate has always been in a state of flux and that the current patterns of change are merely part of these long-standing natural cycles, rather than the result of human activity.

This narrative is appealing for its simplicity and its capacity to absolve responsibility. By framing climate change as a natural phenomenon, beyond the influence or control of human actions, these organizations have effectively stalled significant segments of the public and policy discourse on climate action. The underlying message—of skepticism toward scientific data and distrust of environmental experts—resonates with individuals and groups predisposed to doubting mainstream science or those wary of regulatory interventions in the economy.

The success of this misinformation campaign is not merely in its ability to sow doubt; it lies in its capacity to polarize discussion, turning climate change from a unifying global challenge into a contentious debate mired in political and ideological divisions. This polarization makes it exceedingly difficult to achieve the consensus necessary for comprehensive and collective action on climate issues.

The activities of these denialist groups extend beyond mere advocacy; they actively engage in lobbying efforts aimed at influencing policy decisions and regulatory frameworks. By injecting their narratives into the political bloodstream, they ensure that climate change remains a divisive and contentious issue, often stalemated by partisan divisions and ideological battles.

In examining the machinery of misinformation, it becomes evident that the battle against climate change is not just a scientific or environmental challenge but a deeply political and ideological struggle. Overcoming the inertia of denial and misinformation requires a comprehensive approach, one that addresses not only the scientific and technical aspects of climate change but also the socio-political and economic structures that perpetuate denial and delay action.

Political Polarization

As the machinery of misinformation churned its wheels, climate change denial found fertile ground in the realm of political ideology, particularly within certain segments of the political right. This phenomenon transformed the debate over climate change from a scientific discourse into a deeply polarized ideological battleground. Rejecting the consensus on climate science became more than a denial of empirical evidence; it emerged as a potent symbol of resistance against perceived governmental overreach and an embodiment of the struggle for economic freedom.

This ideological entrenchment of climate change denial has profound implications. It frames the discourse not merely in terms of factual disagreement but as a clash of fundamental values and worldviews. For adherents, disputing the science of climate change aligns with broader themes of individual liberty, skepticism of regulatory interventions, and a preference for market-driven solutions over governmental mandates. Within these circles, questioning climate change transcends scientific skepticism, becoming a marker of political and cultural identity—a badge of defiance against what they perceive as an intrusive, overreaching government intent on imposing its will on the market and individual freedoms.

The consequence of this polarization is a deeply entrenched divide, where attitudes towards climate change are predictive of political affiliations, and vice versa. This divide complicates efforts to address climate change, as the issue is subsumed under broader political and ideological conflicts, making bipartisan support for climate policies increasingly challenging to achieve. The debate over climate change, thus, is not confined to the veracity of scientific claims but extends into fundamental disagreements over governance, economic policy, and the role of the state.

The entanglement of climate change denial with political identity also fosters an environment where facts are often overshadowed by allegiance to party lines. As a result, the discourse around climate policies becomes mired in political maneuvering, with substantive discussions on mitigation and adaptation strategies taking a back seat to ideological battles. This polarization not only stalls progress on climate action but also deepens societal divides, making consensus and collaborative efforts more elusive.

In this charged atmosphere, disentangling climate change from political ideology becomes a crucial challenge. Addressing this issue requires a nuanced approach that respects and navigates these deeply held beliefs and values, seeking common ground that transcends partisan divides. The goal is to reframe climate change as a universal concern that transcends political boundaries, emphasizing shared values and collective responsibility over division and discord. Only through such an inclusive and unifying approach can the global community hope to overcome the obstacles posed by political polarization and move towards effective and equitable climate solutions.

Impact on Policy and Public Opinion

The ripple effects of climate change denial on policy and public opinion have been significant and far-reaching. In nations like the United States, the polarization fostered by denialism has deeply influenced public sentiment on climate change, creating a chasm that complicates the path toward effective environmental policy. This division in public opinion reflects not just differing views on the science of climate change but also on the appropriate response to it, making the enactment of comprehensive climate policies a daunting challenge.

At the core of this challenge is the manner in which climate change has been politicized, turning it into a contentious issue that aligns with broader ideological divides. This politicization means that public support for climate action is often contingent upon political identity, rather than a unified concern for environmental sustainability or the welfare of future generations. Such polarization impedes the development of cohesive climate strategies, as policy proposals become battlegrounds for ideological disputes rather than forums for scientific and economic debate.

The influence of climate change denial extends beyond national borders, casting a long shadow over international efforts to combat global warming. Denialism has played a role in undermining the consensus required for global negotiations and agreements, as nations grappling with internal divisions over climate science are less likely to commit to ambitious international climate initiatives. The result is a fragmented global response to a crisis that demands unified action, delaying critical measures needed to mitigate climate impacts and adapt to a changing world.

The obstruction of climate policy and the erosion of public consensus on climate change are among the most damaging consequences of denialism. These effects are manifest in slowed progress toward renewable energy adoption, inadequate funding for climate resilience and adaptation projects, and the persistence of policies that favor fossil fuel consumption over environmental sustainability. The delay in action exacerbates the challenges posed by climate change, increasing the risk of irreversible damage to ecosystems, economies, and communities worldwide.

Addressing the impact of climate change denial on policy and public opinion requires a concerted effort to bridge divides and build a shared understanding of the urgency of the climate crisis. This endeavor involves not only countering misinformation with facts but also engaging in dialogue that addresses underlying fears and concerns about economic and societal changes. Ultimately, shifting the discourse from one of division to one of collective action and responsibility is crucial for mobilizing the broad-based support necessary for meaningful climate policies and for reinvigorating international cooperation on this critical issue.

The Cost of Inaction

As the Earth continued its relentless march towards higher temperatures, the tangible consequences of climate change denial and inaction manifested with stark clarity. The world bore witness to environmental catastrophes of both increased frequency and ferocity, underscoring the high stakes of ignoring the clarion call for urgent climate action. Wildfires, hurricanes, floods, and heatwaves—each more severe than the last—served as harrowing reminders of the planet's distress.

The wildfires that raged across continents were not only more extensive but also more destructive, obliterating ecosystems, homes, and lives with unprecedented intensity. These infernos, fueled by dry conditions and exacerbated by rising temperatures, underscored the dire predictions of climate scientists, becoming a grim reality that could no longer be ignored.

Hurricanes, too, gained in strength and destructiveness, their enhanced power directly linked to the warming of the oceans. The resulting storms left trails of devastation, upending communities and economies, and prompting a reevaluation of what could be considered "once in a century" events. The increasing frequency and intensity of these storms belief such classifications, signaling a new and more dangerous normal.

Floods washed over regions with merciless abandon, overwhelming defenses and displacing millions. The rising sea levels, coupled with more intense rainfall events, transformed what were once manageable natural occurrences into catastrophic disasters. Cities and countries struggled to adapt to the new realities, finding their existing infrastructure woefully inadequate in the face of such relentless onslaughts.

Heatwaves, too, took a heavy toll, with temperatures soaring to lethal levels in areas unaccustomed to such heat. The health impacts were immediate and devastating, with vulnerable populations suffering the brunt of the consequences. These heat waves served as a stark reminder of the insidious nature of climate change, silently claiming lives and challenging the resilience of communities.

The increasing visibility of these climate-related disasters brought the costs of inaction into sharp relief. The economic toll, measured in billions of dollars in recovery and reconstruction, paled in comparison to the human cost—lives lost, communities shattered, and ecosystems irreparably damaged. Each disaster served as a tragic testament to the folly of denial and delay, highlighting the urgent need for comprehensive and immediate action to mitigate the impacts of climate change.

The cost of inaction is not merely a future concern but a present crisis, demanding immediate attention and action from all sectors of society. It underscores the necessity of transcending denial and division to confront the reality of climate change head-on, with a commitment to implementing solutions that can avert further catastrophe and secure a sustainable future for the planet and its inhabitants.

Facing the Truth

Amidst the escalating climate crisis, a pivotal shift began to emerge. The accumulation of undeniable evidence, coupled with the rising tide of global climate activism, marked a turning point in the battle against climate change denial. The impassioned voices of youth, the wisdom of indigenous communities, and the steadfast dedication of climate scientists converged into a powerful chorus calling for immediate action and accountability. Despite this momentum, the deep-rooted legacy of denial and division continued to cast a long shadow, complicating the path toward a unified and effective response to the climate emergency.

"Ignoring the Inferno" stands as a poignant reflection on the dangers of dismissing scientific consensus and procrastinating in the face of looming existential threats. The challenge before us is daunting, yet the imperative to act has never been more critical. The legacy of denial has left us precariously perched on the brink of irreversible ecological and societal upheaval. However, the growing consensus around the need for urgent, collective action offers a glimmer of hope. This is a moment for humanity to rally together, to leverage our collective ingenuity, compassion, and determination to forge a path forward.

The cost of continued inaction is unimaginable, dwarfing any perceived burdens associated with transitioning to a sustainable future. The stakes are nothing less than the preservation of our planet and the well-being of current and future generations. "Ignoring the Inferno" serves not just as a cautionary tale, but as a galvanizing force, urging us to confront the truth of our situation and to act decisively.

The time to heed the call is now. With each passing moment, the window of opportunity to mitigate the worst impacts of climate change narrows. Our actions today will determine the legacy we leave for the generations that follow. In the face of the greatest challenge humanity has ever encountered, the choice is clear: we must unite, face the truth of our climate crisis, and work tirelessly towards creating a sustainable and just world for all.

Chapter 2: Oceans of Plastic - Marine Pollution Unabated

Marine pollution, prominently marked by plastic waste, stands as a glaring testament to environmental degradation in our times. The oceans, crucial for Earth's ecological balance and human economies, are now battlegrounds against vast expanses of plastic debris. These pollutants span from densely populated coastal waters to the most isolated stretches of the open sea. This chapter aims to dissect the intricate issue of oceanic plastic pollution, tracing its origins, delineating its impact, and exploring comprehensive efforts needed to curb this growing crisis. Through a thorough examination, we seek to unveil the direct link between human actions and the health of our marine environments, emphasizing the urgent need for immediate and enduring interventions.

Origins of Plastic Pollution

In the annals of modern innovation, the advent of plastic in the 20th century marked a turning point. Celebrated for its versatility, durability, and cost-effectiveness, plastic quickly became a cornerstone of industrial manufacturing and a staple in daily consumer life. Its introduction was seen as a herald of convenience and a leap forward in material science, allowing for the production of a myriad of products ranging from disposable packaging to complex components in technology and transportation. However, the very characteristics that made plastic so revolutionary—particularly its durability—soon revealed a darker consequence: a burgeoning environmental crisis.

The escalation of plastic pollution can be traced back to the exponential increase in plastic production coupled with its widespread integration into virtually every aspect of human activity. From packaging materials and single-use products to durable goods and industrial applications, plastic's omnipresence has led to a dramatic rise in disposal rates. Unlike organic materials that biodegrade and reintegrate into the ecosystem, plastics persist in the environment, often for hundreds of years, leading to the accumulation of waste in landfills and, more critically, in marine environments.

The lifecycle of plastic products encapsulates a series of systemic inefficiencies and consumer behaviors that exacerbate the issue. The journey of a plastic item from its creation to its eventual disposal is fraught with points of leakage into the environment. These inefficiencies stem from inadequate waste management systems, limited recycling infrastructure, and a culture of disposability that prioritizes convenience over sustainability. The result is a relentless stream of plastic waste that finds its way into rivers, lakes, and ultimately the oceans, where it poses a severe threat to marine life and ecosystems.

The infiltration of plastics into marine environments is particularly insidious due to their non-biodegradable nature. Once in the ocean, plastics break down into smaller fragments, known as microplastics, which are ingested by marine organisms, entering the food chain and causing a cascade of ecological consequences. These range from the physical entanglement and ingestion of plastic debris by marine animals to the more subtle but equally deadly impacts of chemical leaching and toxicity.

The origins of plastic pollution are deeply rooted in the very fabric of modern society, reflecting a broader issue of unsustainable consumption and production patterns. The journey of plastic from a symbol of modernity to an environmental scourge highlights the need for a systemic reevaluation of how we produce, use, and dispose of materials. Addressing the crisis of plastic pollution requires an approach that encompasses improving waste management practices, fostering innovation in sustainable materials, and cultivating a societal shift towards more responsible consumption habits. As we move forward, understanding the origins of this issue is crucial in charting a path towards a more sustainable interaction with our planet's precious resources.

Current State of Oceanic Plastic Pollution

The vast expanse of the world's oceans, enveloping more than 70% of the Earth's surface, has unwittingly become the final resting place for an ever-increasing deluge of plastic waste. Each year, millions of tons of plastics are swept into marine ecosystems, a testament to the scale and severity of oceanic plastic pollution. The data on the quantity and variety of plastic waste in marine environments paint a grim picture of the situation. It is not just coastal areas that bear witness to this environmental calamity; the remote gyres of the open ocean have become swirling repositories of debris, far from the sight of the casual observer, yet irrefutably present.

The complexity of the issue is magnified by the proliferation of microplastics, tiny fragments resulting from the breakdown of larger plastic items over time. These minuscule particles have infiltrated marine waters at every level, from the sunlit surfaces to the darkened depths, settling into the sediment and even becoming embedded within the ice of the Arctic. The omnipresence of microplastics across diverse marine environments underscores the insidious and pervasive nature of plastic pollution.

This widespread distribution of plastics and microplastics in the ocean disrupts marine life in profound ways. Marine animals, mistaking plastics for food, ingest them, leading to internal blockages, starvation, and often death. The issue extends beyond individual tragedies, affecting entire marine ecosystems. Microplastics, in particular, pose a significant threat as they are ingested by plankton, the foundation of the marine food web, thereby infiltrating and accumulating up the trophic levels, from the smallest fish to apex predators.

Plastics in the ocean are not merely a physical threat. They act as carriers for pollutants, absorbing toxic substances from the water and transferring them to marine organisms upon ingestion. This chemical dimension adds another layer of complexity to the environmental impact of oceanic plastic pollution, affecting not only marine life but potentially human health through the consumption of contaminated seafood.

The current state of oceanic plastic pollution is a stark reminder of the urgent need for global action to mitigate this crisis. The pervasive distribution of plastics across all marine environments, from beaches and coastal waters to the most remote open ocean gyres and Arctic ice, highlights the scale of human impact on the planet's natural systems. Addressing this issue requires concerted efforts to reduce plastic production and consumption, enhance waste management and recycling systems, and foster international collaboration to clean up the legacy of pollution that threatens the health of our oceans and the myriad forms of life they support.

Effects on Marine Species

The insidious nature of plastic pollution extends its reach deep into the marine ecosystem, impacting a wide range of species with devastating consequences. From the smallest zooplankton to the largest marine mammals and seabirds, no creature is immune to the threats posed by the pervasive spread of plastic debris in the oceans. The detrimental effects on marine life are manifold, including ingestion, entanglement, and exposure to toxic chemicals, each contributing to a distressing decline in marine biodiversity and the health of individual species.

One of the most direct impacts of plastic pollution is the ingestion of plastic pieces by marine animals. Mistaking small plastic particles for food, organisms across the marine food web consume these lethal fragments. This ingestion can lead to severe internal blockages, disrupting digestive processes and often leading to starvation and death. Moreover, plastics are known to leach toxic substances, such as bisphenol A (BPA) and phthalates, which can lead to poisoning and various health issues in marine life. Species as varied as tiny zooplankton, fish, sea turtles, and seabirds have been documented ingesting plastics, illustrating the widespread nature of this threat.

Beyond ingestion, marine species also face the danger of entanglement in plastic waste. Abandoned fishing gear, known as "ghost nets," along with other forms of plastic waste, create deadly traps for a wide array of marine animals. Entanglement can lead to physical injuries, infections, impaired mobility, starvation, drowning, or suffocation. Large mammals like whales and dolphins, as well as sea turtles and seabirds, are particularly vulnerable to entanglement, which can lead to prolonged suffering and eventual death.

The severity of plastic pollution's impact on marine species is underscored through various case studies. For instance, autopsies of seabirds have revealed stomachs filled with plastic debris, leading to malnutrition and starvation. Similarly, whale strandings have occasionally been linked to the ingestion of large amounts of plastic, causing internal blockages and death. These case studies are not isolated incidents but rather indicators of a much larger issue threatening marine biodiversity at a global scale.

The effects of plastic pollution on marine species are a stark reminder of the ecological crisis unfolding in our oceans. The lethal consequences of ingestion and entanglement, coupled with the toxic threat posed by chemicals associated with plastics, underscore the urgent need for comprehensive strategies to address plastic pollution. Protecting marine biodiversity and ensuring the survival of individual species demand immediate action to mitigate the influx of plastics into marine environments and to clean up the existing pollution. Only through concerted global efforts can we hope to reverse the tide of plastic pollution and safeguard the future of marine life.

Disruption of Marine Ecosystems

The menace of plastic pollution extends far beyond the immediate threats posed to individual marine species, casting a long shadow over the integrity and vitality of entire marine ecosystems. Central to this issue is the pervasive infiltration of microplastics into marine food webs, a phenomenon that encapsulates the profound and potentially irreversible impacts on ecosystem health and functionality. These tiny particles, less than five millimeters in diameter, have been found in the remotest marine environments, from the surface waters to the deep-sea sediments, indicating the widespread and entrenched nature of the problem.

Microplastics serve as a vector for toxicity within marine ecosystems, absorbing and concentrating pollutants from the surrounding water before entering the food chain. Once ingested by plankton—the foundational layer of the marine food web—these pollutants can bioaccumulate, transferring from prey to predator and magnifying in concentration up the food chain. This process poses significant, yet largely unknown, long-term threats to the health and survival of marine species, from the smallest organisms to top predators.

The physical presence of plastic debris in marine environments disrupts the natural habitat, altering the physical landscape in which marine species have evolved to thrive. Large accumulations of plastic can smother seabeds, choke coral reefs, and obstruct the movements of animals, affecting species composition and abundance. Such alterations to the habitat can tip the ecological balance, leading to shifts in biodiversity and ecosystem function that may prove difficult, if not impossible, to reverse.

The disruptive impact of plastic pollution on marine ecosystems is a clarion call for urgent action. It underscores the need for a comprehensive understanding of the ecological roles and interactions within these systems to gauge the full extent of the damage. As plastics continue to accumulate in our oceans, the specter of a looming ecological shift looms large, threatening not only the diversity and productivity of marine life but also the goods and services these ecosystems provide to humanity. Addressing this crisis requires an approach that, encompassing not only the reduction of plastic waste but also the restoration and preservation of marine habitats, to safeguard the health and resilience of our oceans for future generations.

Human Health Risks

The pervasive spread of plastics into marine ecosystems carries with it not only environmental consequences but also significant implications for human health. Microplastics, which have become ubiquitous in marine environments, pose a particularly insidious risk. These minute particles are known for their ability to absorb and concentrate toxic chemicals from seawater, such as polychlorinated biphenyls (PCBs), dioxins, and other pollutants that are harmful to human health. As these plastics enter the food chain, they make their way onto our plates, primarily through seafood, introducing these concentrated toxins into the human body.

Emerging research into the health risks associated with microplastic ingestion is raising alarms. One of the primary concerns is the potential for endocrine disruption, given that many of the chemicals that adhere to microplastics are known endocrine disruptors. These substances can interfere with hormone systems, potentially leading to a variety of health issues, including reproductive abnormalities, neurological problems, and increased cancer risk. Moreover, the physical presence of microplastics in the human digestive system is a subject of ongoing investigation, with researchers exploring the possibility that these particles could cause inflammation or even transfer toxins across the gut barrier.

The true extent of the health risks posed by microplastics is still coming into focus, with many questions remaining unanswered. The potential for bioaccumulation—whereby concentrations of toxic substances increase in the body over time—adds another layer of concern, suggesting that even minimal exposure, if persistent, could have long-term health implications. This uncertainty underscores the critical need for comprehensive and rigorous scientific studies to fully understand the implications of microplastic ingestion on human health.

The issue of microplastics and human health is not just a matter of individual exposure but reflects broader concerns about the safety and sustainability of our food systems and the health of marine environments. It highlights the interconnectedness of human health and environmental health, underscoring the need for a holistic approach to addressing plastic pollution. As the body of evidence grows, it becomes increasingly clear that tackling the spread of plastics in our oceans is not only an ecological imperative but a public health priority. Addressing this challenge requires concerted action across sectors and disciplines, from enhancing waste management and reducing plastic production to supporting research and public health initiatives aimed at understanding and mitigating the impacts of plastic pollution on human health.

Economic Costs and Losses

The ripple effects of marine plastic pollution extend far beyond environmental degradation, casting a shadow over the economy as well. Industries that depend on vibrant and healthy marine ecosystems face substantial economic challenges as a result of this pervasive issue. Notably, the tourism and fisheries sectors, which are cornerstone industries for many coastal economies, bear the brunt of these economic damages.

The tourism industry suffers significantly from marine plastic pollution. Pristine beaches and clear waters are among the primary attractions for coastal destinations. However, the accumulation of plastic waste not only mars the natural beauty of these locales but also poses health risks to visitors, ultimately deterring tourism. The visual and physical pollution undermines the appeal of beach destinations, leading to a decline in visitor numbers and, consequently, a reduction in income for local businesses and communities that rely on tourism. The cost of cleaning up beaches to restore their appeal further strains local economies, diverting resources that could be used for community development or conservation efforts.

Similarly, the fisheries sector faces direct economic repercussions from plastic pollution. Contaminated fish stocks, resulting from the ingestion of plastics and associated toxins by marine life, affect the quality and safety of seafood. This contamination can lead to declines in fish populations, reduced catches for fishermen, and potential health warnings against consuming certain seafood, all of which contribute to economic losses for the fishing industry.

Additionally, the entanglement of marine animals in plastic debris and abandoned fishing gear—a phenomenon known as "ghost fishing"—results in further declines in fish stocks and biodiversity, exacerbating the economic challenges faced by the sector.

Beyond the direct impacts on tourism and fisheries, the broader economic burden of managing and cleaning up marine plastic pollution is substantial. Municipalities and governments spend millions of dollars annually on waste collection, management, and cleanup efforts. This financial strain underscores the economic rationale for investing in preventive measures, such as reducing plastic usage and improving waste management infrastructure. Implementing more efficient and sustainable waste management systems not only mitigates the environmental impacts of plastic pollution but also represents a cost-effective approach in the long term.

The economic costs and losses associated with marine plastic pollution highlight the urgent need for collective action and policy interventions. Investing in solutions that prevent plastic waste from entering marine ecosystems, along with efforts to clean up existing pollution, is essential for protecting the health of marine environments and ensuring the economic viability of industries that depend on them. Addressing this challenge requires a strategy that involves stakeholders at all levels, from local communities to international organizations, working together to safeguard our oceans and the economies they support.

Global Initiatives and Policies

Addressing the formidable challenge of marine plastic pollution requires a coordinated global response, underpinned by effective policymaking and international collaboration. Recognizing the transboundary nature of plastic waste and its ability to impact marine ecosystems far from its point of origin, a variety of global initiatives and policies have been put forth to stem the tide of plastic entering our oceans.

One notable example is the United Nations' Clean Seas campaign, launched with the aim of galvanizing governments, the public, and the private sector to take definitive action against marine plastic pollution. This initiative represents a concerted effort to raise awareness, promote best practices, and encourage the implementation of strategies to reduce plastic waste. By advocating for the reduction of single-use plastics and encouraging changes in consumer behavior, the Clean Seas campaign seeks to address plastic pollution at its source.

International commitments, such as those made under the framework of the United Nations Environment Programme (UNEP), further highlight the global consensus on the need to confront this issue head-on. Countries around the world have pledged to take action to prevent plastic waste from entering the marine environment, through measures ranging from improving waste management infrastructure to instituting bans on certain types of single-use plastics.

At the national level, policies and regulations play a pivotal role in curbing plastic pollution. Many countries have introduced legislation aimed at reducing plastic production and consumption, enhancing recycling rates, and improving waste management practices. These include plastic bag bans, requirements for packaging to be recyclable or compostable, and incentives for businesses and consumers to adopt sustainable practices.

In addition to regulatory measures, there is a growing emphasis on the development of circular economies, where the design, production, and use of materials are optimized to minimize waste and promote the reuse and recycling of resources. This approach not only addresses the issue of plastic pollution but also offers a path toward more sustainable economic and environmental practices.

The global fight against marine plastic pollution is bolstered by the participation of non-governmental organizations (NGOs), research institutions, and the private sector, all of which contribute to the development and implementation of innovative solutions. From advancements in biodegradable materials to initiatives aimed at cleaning up existing marine plastic, these collaborative efforts underscore the approach required to tackle this crisis effectively.

The emergence of global initiatives and policies dedicated to combating marine plastic pollution underscores the recognition of its severity and the commitment to finding solutions. While challenges remain, the ongoing efforts at international, national, and local levels offer hope for a future in which the oceans are free from the scourge of plastic waste.

Innovative Solutions and Community Actions

In the campaign against marine plastic pollution, the front lines are occupied not only by policymakers and global initiatives but also by a wave of innovation, community engagement, and heightened public consciousness. The battle is increasingly characterized by technological breakthroughs in recycling, the emergence of alternative materials that could diminish our reliance on conventional plastics, and successful operations aimed at removing existing waste from our oceans. These achievements are complemented by community-driven efforts and grassroots movements, showcasing the power of collective action in forging a path toward sustainability and environmental stewardship.

Technological advancements in plastic recycling are revolutionizing the way we approach plastic waste, making it possible to recover and repurpose materials that would otherwise contribute to pollution. Innovations such as enhanced sorting technologies, chemical recycling methods that break plastics down to their molecular components, and upcycling processes that transform waste into higher-value products, are critical in closing the loop on plastic use and reducing the volume of waste that enters the marine environment.

Simultaneously, the development of alternative materials represents a proactive approach to reducing plastic pollution at its source. Biodegradable plastics, made from natural substances such as plant starches, and innovations in packaging design that eliminate the need for plastic altogether, are gaining traction as viable solutions. These materials offer the promise of a future where products and packaging break down harmlessly in the environment, circumventing the long-term ecological impacts associated with traditional plastics.

On the ground, the success of cleanup operations in coastal areas and open waters highlights the direct impact that concerted action can have on reducing marine plastic pollution. Organizations and initiatives, ranging from local beach cleanups to ambitious projects targeting oceanic garbage patches, demonstrate the tangible benefits of removing plastics from marine environments. These efforts not only mitigate the immediate threat to marine life but also raise public awareness of the scale of the problem and the importance of preventing further pollution.

Community-driven efforts and grassroots movements are the lifeblood of the campaign against plastic pollution, embodying the principle that meaningful change often begins at the local level. Through education, advocacy, and action, communities around the world are making significant strides in reducing plastic waste. Initiatives such as plastic-free challenges, community recycling programs, and educational campaigns that promote sustainable living practices empower individuals to make choices that benefit the environment.

The fight against marine plastic pollution is bolstered by a combination of innovative solutions and community action, illustrating the approach required to address this global challenge. As technological advancements continue to emerge and public awareness grows, the potential for significant reductions in plastic waste and a shift toward a culture of environmental responsibility becomes ever more apparent. Through collective effort and a commitment to innovation, the goal of restoring the health and vitality of our oceans moves closer to reality.

Chapter 3: Choking Skies - The Neglect of Air Quality

As the dawn breaks, a hazy sun rises above the horizons of bustling cities around the globe, casting an eerie, murky glow. This dim light unveils a grim tableau — a pervasive, insidious crisis that looms over humanity, largely unnoticed and gravely neglected: air pollution. This chapter looks into the shadowy depths of deteriorating air quality, an unseen assassin that silently afflicts billions, navigating through its origins, impacts, and the collective failure to confront this menace head-on.

Breathing in the Problem

The deterioration of air quality, a pressing concern shadowing our planet, has deep roots stretching back to the dawn of the Industrial Revolution. It was then that humanity first stoked the fires of large-scale industry, inadvertently igniting the enduring crisis of air pollution. However, it's the explosive industrialization and urban sprawl of recent decades that have pushed this issue to the brink of criticality. At the heart of this environmental onslaught lie several key perpetrators: vehicular emissions, industrial outputs, the billowing dust of construction sites, and the smoldering remnants of agricultural burn-offs. These sources spew a cocktail of toxic particulates and gases into the air, weaving a dense veil of pollution that hangs over cities and countrysides alike.

Despite mounting scientific evidence linking air pollution to a host of maladies — from respiratory and cardiovascular diseases to stunted cognitive development and diminished ecosystem vitality — the response from regulatory bodies and industry stakeholders has been frustratingly tepid. The clear and present danger posed by air pollutants is well-documented and universally recognized. Yet, the push for substantive regulatory and remedial action has been mired in a quagmire of economic considerations, political inertia, and a concerning lack of global solidarity.

This lackadaisical approach to addressing air quality degradation not only undermines efforts to safeguard public health but also betrays a gross undervaluation of the environment's critical role in human prosperity. The consequences of such negligence are far-reaching, impacting billions of lives and threatening the very fabric of natural ecosystems worldwide. As we grapple with the complexities of mitigating air pollution, the clock continues to tick, underscoring the urgent need for a collective, concerted effort to clear the air. Without significant changes in policy, technology, and public awareness, the specter of choking skies will continue to loom large over our shared future, a grim reminder of what happens when we fail to breathe life into the solutions that lie within our grasp.

The Invisible Killers

Within the haze that blankets our cities and countrysides, a silent but deadly arsenal of pollutants wages war on human health and the natural environment. Among the most formidable of these adversaries are particulate matter (PM), nitrogen oxides (NOx), sulfur dioxide (SO2), and volatile organic compounds (VOCs). Invisible to the naked eye, these pollutants carry with them a legacy of destruction, insidiously undermining the vitality of ecosystems and posing grave risks to human health. Their origins are as varied as their effects are dire, stemming primarily from the combustion processes that power our vehicles, energize our industries, and clear our agricultural lands.

Once liberated into the atmosphere, these pollutants embark on a relentless assault against clean air. They degrade the quality of the very air we breathe, not only through their direct presence but also by playing a critical role in the formation of ground-level ozone and secondary particulates. These secondary pollutants add another layer of complexity to the air quality crisis, further exacerbating the challenges we face in mitigating the impact of air pollution.

The chemistry of destruction that unfolds in our atmosphere is both complex and catastrophic. Nitrogen oxides and volatile organic compounds, under the influence of sunlight, participate in a series of reactions that give birth to ground-level ozone, a component of smog that is notorious for its ability to cause respiratory problems and damage vegetation. Particulate matter, depending on its size, can penetrate deep into the lungs and even enter the bloodstream, causing a range of health issues from heart attacks to aggravated asthma. Sulfur dioxide, a byproduct of burning fossil fuels, contributes to the acidification of rain, harming aquatic life and vegetation, and eroding buildings and historical monuments.

The invisibility of these pollutants belies their potency. They are killers on the loose, silent assassins that claim millions of lives each year and wreak havoc on the natural world. As the evidence of their impacts mounts, the imperative for action grows stronger. Addressing the sources of these pollutants, transitioning to cleaner energy sources, and enhancing regulatory frameworks are crucial steps in our battle to reclaim the air. The fight against these invisible killers is a fight for our health, our environment, and our future. It's a battle we cannot afford to lose.

A Toll on Human Health

The toll that air pollution exacts on human health is both profound and pervasive, insidiously infiltrating the lives of billions across the globe. The adverse health effects span a broad spectrum, affecting nearly every system within the human body. In the short term, exposure to polluted air can trigger and exacerbate respiratory conditions, such as asthma and bronchitis, manifesting in wheezing, coughing, and shortness of breath. These immediate reactions, though alarming, barely scratch the surface of the potential long-term consequences.

Chronic exposure to air pollutants has been incontrovertibly linked to a host of severe health issues. Heart disease, stroke, lung cancer, and chronic obstructive pulmonary disease (COPD) have all been associated with prolonged exposure to the noxious blend of particulates and gases that constitute polluted air. Moreover, there is a growing body of evidence suggesting that air pollution may also contribute to the development of diabetes, dementia, and low birth weights, further underscoring the insidious nature of this threat.

Perhaps the most sobering statistic comes from the World Health Organization (WHO), which estimates that millions of premature deaths occur annually as a direct result of air pollution. This staggering figure highlights not just the scale of the crisis but also its acute lethality. The impact of air pollution on health is not just a matter of morbidity but mortality, shortening life spans and diminishing the quality of life for countless individuals.

Compounding this tragedy is the unequal burden of air pollution. Children, the elderly, and residents of low-income communities bear the brunt of this crisis. These vulnerable populations, due to their physiological, socioeconomic, and geographical circumstances, are often the most exposed to polluted air and, consequently, the most affected by its deleterious health effects. The disparity in exposure and impact underscores a grave injustice, one that amplifies the urgency of addressing air pollution not only as an environmental issue but as a pressing public health and social equity challenge.

Environmental and Economic Fallout

The insidious reach of air pollution extends far beyond its immediate impact on human health, casting a long shadow over the natural environment and the global economy. The environmental degradation wrought by air pollutants is both diverse and profound, affecting nearly every aspect of the natural world.

One of the most visible manifestations of this environmental damage is acid rain. This phenomenon, primarily caused by the emissions of sulfur dioxide (SO_2) and nitrogen oxides (NO_x), precipitates a cascade of harmful effects on forests, soils, and aquatic ecosystems. Acid rain leaches essential nutrients from the soil, weakens trees by damaging their leaves, and acidifies lakes and streams, disrupting the delicate balance of aquatic life. The repercussions of these changes ripple through ecosystems, leading to diminished biodiversity and the disruption of habitats.

Air pollution poses a significant threat to agricultural productivity. Pollutants like ozone can impair pollination and stunt the growth of crops, directly impacting food security. This reduction in yield not only threatens the livelihood of farmers but also contributes to global food scarcity challenges, highlighting the interconnectivity of environmental health and human wellbeing.

The economic repercussions of air pollution are equally staggering. The healthcare costs associated with treating conditions related to air pollution are astronomical, burdening public health systems and draining resources that could be invested elsewhere. Additionally, the lost labor productivity due to illness and premature mortality translates into billions of dollars in economic losses each year. These financial strains underscore the economic inefficiency of inaction against air pollution.

The environmental and economic fallout from air pollution paints a clear picture: the cost of inaction far exceeds the cost of addressing the issue head-on. Investing in cleaner technologies, implementing stringent regulations, and fostering global cooperation to combat air pollution are not just environmental imperatives but economic necessities. By taking decisive action, societies can safeguard not only the health of their citizens but also the resilience of their economies and the vitality of the natural world.

The Global Response: A Patchwork of Efforts

In the face of the clear and present danger posed by air pollution, the global response has been a patchwork of efforts, varying significantly in scope, scale, and effectiveness. While some countries have stepped up to the challenge, implementing robust regulations and pioneering cleaner technologies, others have struggled to make meaningful progress, their efforts stymied by economic constraints, political apathy, or both.

The disparity in national responses reflects a broader inconsistency that undermines global initiatives to combat air pollution. Some nations, recognizing the dire health and environmental implications of unchecked emissions, have adopted stringent standards for air quality, invested in renewable energy sources, and incentivized the adoption of electric vehicles. These measures, though effective, highlight the unevenness of global efforts, with proactive policies often concentrated in wealthier countries possessing the resources to enact them.

On the international stage, numerous agreements and guidelines aim to unify global efforts to tackle air pollution. The World Health Organization (WHO) and other international bodies have set forth air quality standards, hoping to steer countries toward cleaner, healthier environments. However, the enforcement of these guidelines presents a formidable challenge. Without binding legal authority or mechanisms to compel compliance, international standards often serve more as aspirational benchmarks than actionable mandates.

The struggle to meet even the basic air quality standards laid out by health organizations is a testament to the complexities of managing air pollution on a global scale. Developing countries, in particular, face a daunting task: balancing the imperative of economic growth with the necessity of environmental stewardship. The pressure to industrialize and urbanize can push environmental concerns to the backburner, exacerbating pollution levels and widening the gap in global air quality.

The current state of global air quality and the efforts to improve it underscore a critical need for increased cooperation, enhanced support for developing nations, and a concerted push for innovation in pollution management technologies. Without a unified, determined response to air pollution, the patchwork of efforts will remain inadequate, leaving populations vulnerable and ecosystems at risk. The challenge is not insurmountable, but overcoming it requires a collective acknowledgment that the health of our planet and its inhabitants is an investment worth making, one that demands action, commitment, and shared responsibility from all corners of the globe.

Innovations and Solutions

In the midst of the grim landscape painted by the crisis of air pollution, "Choking Skies" casts a hopeful light on the innovative strides being made towards cleaner air and a healthier planet. At the forefront of these advancements are renewable energy technologies, electric vehicles, and rigorous emissions standards — each representing a critical piece in the puzzle of air quality improvement. These innovations signal a shift away from fossil fuel dependency, a major contributor to air pollution, towards more sustainable and environmentally friendly alternatives.

Urban planning also plays a pivotal role in the blueprint for cleaner air. Initiatives aimed at promoting public transport, cycling, and walking not only reduce reliance on personal vehicles but also cut down on vehicular emissions, one of the primary sources of urban air pollution. Furthermore, the integration of green spaces within urban landscapes offers a natural means of filtering pollutants, providing urban dwellers with much-needed oases of cleaner air.

As "Choking Skies" draws to a close, it issues a compelling call to action, framing the battle against air pollution not merely as an environmental challenge but as a critical issue of health and social equity. The chapter champions a multilayered approach, one that marries stringent regulatory frameworks with the embrace of technological innovation and the power of community activism. This battle cry underscores the non-negotiable urgency of the task at hand: securing clean air transcends the realm of luxury and stands as an indispensable human right, vital for the sustainability of our ecosystems and the health of future generations.

The narrative underscores that the journey towards a pollution-free world is not one to be walked alone but requires a concerted, global effort, infused with political determination and a radical reevaluation of our relationship with the environment. The path ahead is clear — it demands collective action, unwavering political resolve, and an ambitious reimagining of societal values and practices. "Choking Skies" leaves readers with a resonant message: the time for change is now, and the responsibility to act rests on the shoulders of us all.

Chapter 4 Silent Extinctions - Indifference Towards Biodiversity

In the heart of Earth's myriad ecosystems, a silent crisis unfolds, shadowed by the cacophony of human advancement. The planet's biodiversity, a rich tapestry woven over billions of years, is unraveling thread by thread, a phenomenon met with widespread human indifference. This chapter, "Silent Extinctions - Indifference Towards Biodiversity," looks into the nuanced and complex relationship between humanity and the natural world, exploring how our neglect has accelerated the loss of biodiversity and why this matters.

Biodiversity encompasses the variety of all life forms on Earth, including plants, animals, fungi, and microorganisms, as well as the ecosystems they comprise. It stands as the cornerstone of ecosystem services that support life as we know it, playing a crucial role in providing essential resources such as food and water. Beyond these tangible benefits, biodiversity is instrumental in disease management, climate regulation, and even contributes to spiritual fulfillment, underscoring the myriad ways in which the natural world is interwoven with human well-being.

Despite its intrinsic value and critical role in sustaining life, the global community's response to the escalating crisis of biodiversity loss has been tepid at best. This lackluster response is marked by a disconcerting disconnect between the scientific urgency conveyed by researchers and the level of action or concern demonstrated by society at large. The alarming rate at which species are disappearing reflects not just a failure to recognize the significance of biodiversity but also a broader societal indifference towards the mechanisms that underpin our very existence. This gap between understanding and action highlights a critical challenge in addressing biodiversity loss: bridging the divide between scientific knowledge and meaningful societal engagement.

The indifference towards biodiversity loss can be attributed to several factors, chief among them the invisibility of its decline to the everyday observer. Unlike the immediate, tangible impacts of environmental disasters, which often command global attention and prompt quick responses, the extinction of a species tends to occur out of the public eye and without fanfare. This gradual and often unnoticed erosion of biodiversity fails to evoke the urgent reactions that are mobilized in response to more visible environmental crises.

This invisibility is further compounded by a widespread misconception regarding nature's resilience. Many hold the belief that no matter how much we take from the Earth, it possesses an infinite capacity to replenish itself. This erroneous assumption leads to a gross underestimation of the impact human activities have on the planet's ecosystems. The truth, however, is far more complex and concerning; ecosystems have tipping points and thresholds that, when exceeded, can lead to irreversible damage and loss. The belief in nature's boundless resilience contributes significantly to the indifference observed towards the silent, yet accelerating, loss of biodiversity, underscoring the need for a shift in perception and an increase in conservation efforts.

Yet, the reality of our impact on the natural world is starkly different from the misconception of its infinite resilience. Human activities have pushed the Earth's ecosystems to their limits, resulting in a series of destructive effects that threaten the very fabric of biodiversity. Habitat destruction, a primary driver of biodiversity loss, is accelerated by deforestation, urban expansion, and agricultural intensification. These activities fragment and erase the complex habitats species depend on for survival, leaving them with nowhere to turn.

Pollution further compounds the crisis. The chemicals that lace our rivers and the microplastics filling our oceans poison a wide array of life forms, disrupting ecosystems and contributing to the decline of species populations. Overexploitation of natural resources, whether it be through unsustainable fishing, hunting, or logging, strips the natural world of its assets faster than they can be replenished.

Climate change and the introduction of invasive species further exacerbate the pressures on biodiversity. The changing climate alters habitats and conditions so rapidly that many species cannot adapt in time, leading to their decline or extinction. Similarly, invasive species introduced into new environments often outcompete native species for resources, disrupting established ecological balances.

The cumulative effect of these human actions has led our planet into its sixth mass extinction event, this time driven not by natural processes but by humans themselves. This unprecedented rate of loss signifies not just a crisis for the natural world but a profound challenge for humanity, which depends on the diversity of life for its own survival and well-being.

The consequences of biodiversity loss ripple across multiple dimensions, extending far beyond the immediate environmental impacts to touch upon the very foundations of human society. Food security, health, and the global economy are intricately linked to the health of ecosystems, all of which suffer in the wake of biodiversity decline. Ecosystems weakened by the loss of species become less resilient to changes, particularly those induced by climate change. This reduced resilience undermines their ability to provide the services upon which human life depends, including the provision of clean air, water, and a stable climate, as well as the pollination of crops and control of pests.

Biodiversity is not merely a resource for physical sustenance; it enriches our world in ways that transcend material benefits. The loss of biodiversity impoverishes our planet culturally and spiritually, severing the deep connections many communities have with their natural environments. These connections, often forged over millennia, are not only integral to the identity and heritage of these communities but also to the collective cultural wealth of humanity. Indigenous peoples and local communities, in particular, who are the most attuned to their environments, face the loss of traditional knowledge, customs, and languages as the ecosystems they depend on and are part of, deteriorate.

The erosion of biodiversity, therefore, is not just an environmental issue but an immediate crisis that threatens food security, undermines human health, destabilizes economies, and erases cultural identities. Addressing this loss requires a recognition of its broad and profound implications for humanity and a concerted, global effort to safeguard the natural world that sustains us all.

Despite the dire situation biodiversity faces, the global response to its decline has been notably muted. This indifference can be attributed to several key factors, chief among them a widespread lack of awareness and understanding regarding the complexity of biodiversity and its critical importance to human well-being. Many people remain unaware of how intertwined their lives are with the health of ecosystems around the globe, and how the loss of biodiversity directly impacts food security, health, economic stability, and cultural heritage.

Compounding this issue are the economic systems and structures that are major drivers of biodiversity loss. These systems are deeply entrenched, supported by policies and practices that prioritize short-term economic gains over the long-term sustainability of our planet's ecosystems. Such an approach to development and progress has led to unsustainable exploitation of natural resources, habitat destruction, and a continuous erosion of the natural world.

Breaking this cycle of biodiversity loss requires a multi-faceted approach. Raising awareness about the importance of biodiversity and the consequences of its decline is crucial. This involves not only educating the public but also ensuring that leaders and policymakers are informed about the latest scientific findings related to biodiversity. However, awareness alone is insufficient.

There must be a concerted effort to integrate biodiversity considerations into all levels of decision-making, from local community initiatives to international agreements. This includes reevaluating and adjusting the economic models and policies that have contributed to biodiversity loss, aiming for a balance between development and conservation. Shifting towards a more sustainable model of development necessitates innovative thinking and solutions that recognize the value of biodiversity, promoting practices that protect and restore ecosystems rather than deplete them. Only through such comprehensive and integrated approaches can the global community hope to stem the tide of biodiversity loss and move towards a more sustainable and equitable future for all inhabitants of our planet.

Engaging society in the conservation of biodiversity, while a formidable challenge, presents a significant opportunity to alter the trajectory of environmental decline. Education and outreach stand at the forefront of strategies designed to shift public perceptions, moving from indifference to a deep-seated sense of stewardship and responsibility towards the natural world. By increasing awareness of the intricate connections between human well-being and biodiversity, individuals can be motivated to adopt practices that contribute to conservation efforts.

Community-based conservation efforts exemplify the potential of localized, inclusive approaches to protecting biodiversity. These initiatives leverage the unique knowledge, values, and investment of local communities to safeguard ecosystems and the myriad species they support. By involving communities directly in the conservation process, these efforts not only achieve more sustainable outcomes but also empower communities, fostering a sense of ownership and responsibility towards their natural heritage.

The advent of technology and the rise of citizen science initiatives have revolutionized the field of conservation. These tools offer new and innovative ways to monitor biodiversity, providing data and insights that were previously out of reach. Citizen science, in particular, engages the public directly in conservation research, allowing individuals to contribute to scientific knowledge through activities like species monitoring and habitat observation. This direct involvement demystifies scientific research, making it accessible and engaging for people from all walks of life.

These technological advances and participatory approaches not only enhance our understanding and capacity to protect biodiversity but also serve to bridge the gap between scientific communities and the general public. By fostering a more informed and engaged society, we can cultivate a collective ethic of care for the natural world, driving forward conservation efforts with renewed vigor and hope.

As "Silent Extinctions" poignantly illustrates, the widespread indifference towards biodiversity loss transcends environmental concern, mirroring deeper societal values and priorities. This crisis beckons for a collective introspection and reevaluation of humanity's relationship with the natural world, urging us to acknowledge that biodiversity is not merely a luxury but an essential cornerstone for our survival.

In calling for this awakening, "Silent Extinctions" challenges its readers to envision a future where humanity and biodiversity coexist in harmony. It prompts us to imagine a world where our actions are not driven by exploitation but by a mutual respect and understanding, recognizing that our fates are intertwined within the intricate web of life. The chapter argues that thriving together is not only possible but necessary for the sustenance of life on our planet.

Despite the formidable challenges that lie in the path of reversing biodiversity loss, the narrative is imbued with hope. Across the globe, a movement is growing — individuals, communities, and nations are stepping up to protect and restore the natural world. Efforts ranging from the rewilding of landscapes to the implementation of robust conservation laws exemplify the myriad ways in which we can contribute to a more biodiverse, resilient, and sustainable future.

"Silent Extinctions" transcends a mere call to action; it serves as a profound reminder of the resilience and vitality of life on Earth. It celebrates the power of human ingenuity and compassion, showcasing how, through collective will and effort, we can safeguard the planet's biodiversity for future generations. This chapter is not just a reflection on what we stand to lose but a testament to what we can achieve when we come together to protect the very essence of life itself.

Part II: Societal Failures

Chapter 5: The Wealth Chasm - Economic Inequality Explored

In the shadow of towering skyscrapers, the stark contrast between the affluence of the few and the poverty of the many paints a vivid picture of the contemporary world. This disparity, a gaping chasm rather than a mere gap, underlines a profound economic inequality that permeates every facet of society. It's a phenomenon that not only questions the integrity of our economic systems but also challenges the very foundation of our moral and ethical values.

The Mechanics of the Wealth Chasm

At the core of economic inequality lies a sophisticated structured system, an engine of disparity that is propelled by a combination of policies, the mechanisms of global capitalism, and deep-rooted systemic biases. This engine is not a mere assembly of impartial economic principles; rather, it is a construct designed to facilitate and perpetuate the accumulation of wealth among a select segment of society. The dynamics of this system reveal a disturbing trend of increasing wealth concentration, which is manifested through several key mechanisms:

1. Wage Stagnation vs. Soaring Executive Compensations: A striking feature of the contemporary economic landscape is the stark contrast between the stagnation of wages for the majority and the exponential increase in compensation for executives and top-tier professionals. This phenomenon is not confined to isolated sectors but is prevalent across a broad spectrum of industries, highlighting a systemic skew in the distribution of wealth. The disparity in income growth exacerbates the economic divide, relegating a significant portion of the workforce to a cycle of financial precarity.

2. Tax Systems and Loopholes: The architecture of tax systems plays a pivotal role in deepening economic inequality. Crafted with intricacies that favor the affluent, these systems offer loopholes and preferential rates on capital gains, thereby enabling the wealthy to amass greater fortunes while minimizing their tax liabilities. This structural bias not only widens the wealth gap but also undermines the principle of fiscal equity, as the burden of taxation disproportionately falls on those least equipped to bear it.

3. Inheritance and Intergenerational Wealth Transfer: Economic inequality is not merely a phenomenon of the present but a legacy

that is passed down through generations. Inheritance practices and the mechanisms of wealth transfer cement the foundations of disparity, ensuring that affluence remains concentrated within certain families or groups. This cyclical perpetuation of wealth ensures that inequality is deeply entrenched within the social fabric, creating barriers to social mobility and perpetuating a class divide.

The narrative of economic inequality is further complicated by the dual forces of globalization and technological disruption, each playing a significant role in shaping the modern economic landscape:

1. Globalization's Polarizing Effects: While globalization has been lauded for fostering market integration and facilitating cross-border economic activities, it has also led to a polarized economic environment. The migration of manufacturing jobs to regions with lower labor costs has resulted in economic displacement for many, stripping away the livelihoods of workers in more developed economies. This redistribution of jobs, driven by the pursuit of cost efficiency, has left vast segments of the population grappling with economic instability and marginalization.

2. The Rise of Automation and Artificial Intelligence: The rapid advancement of technology, characterized by the increasing adoption of automation and artificial intelligence, marks a significant shift in the economic paradigm. This transition towards mechanized labor has the potential to sideline human workers, concentrating economic power and wealth in the hands of those who control the technologies. The implications of this shift are profound, raising existential questions about the future of work, the distribution of wealth, and the very nature of economic participation in an increasingly automated world.

The exploration of the mechanics behind the wealth chasm sheds light on the complex interplay of factors that contribute to economic inequality. Understanding these dynamics is crucial for devising strategies aimed at bridging the divide, fostering a more equitable distribution of wealth, and ensuring a more inclusive economic future for all.

Manifestations of Economic Inequality

The impact of economic inequality transcends the mere disparity in financial resources, casting long shadows over various aspects of society. It manifests in ways that deeply affect the social fabric, health standards, and educational opportunities, creating a divided landscape where the divide is not only economic but also geographical, cultural, and existential.

1. Social Fragmentation: Economic inequality leads to a societal segmentation, where communities are increasingly divided along the lines of wealth. This division is not just metaphorical; it is physically visible in the urban landscape. The affluent retreat into gated communities and exclusive neighborhoods, enjoying a plethora of resources and services. In stark contrast, less affluent communities face the challenges of underfunded schools, limited access to essential services, and higher rates of crime and violence. This segregation reinforces social barriers, diminishing the sense of community and mutual understanding across economic divides.

2. Health Disparities: Perhaps one of the most direct consequences of economic inequality is seen in the realm of healthcare. Access to quality health services, nutritious food, and healthy living conditions often correlates directly with one's economic standing. The result is a pronounced health divide, where the wealthy enjoy longer lifespans and better overall health, while the less fortunate face higher rates of chronic diseases, mental health issues, and lower life expectancy. This disparity is not merely a matter of individual health outcomes but reflects a broader societal failure to provide for the well-being of all its members.

3. Educational Inequities: Education, often hailed as the great equalizer, becomes a battleground of inequality. As the cost of higher education spirals upwards, it increasingly becomes a

privilege reserved for those with the means to afford it. This financial barrier to education perpetuates a cycle of inequality, where access to knowledge and the opportunities it affords is gatekept by economic status. The ripple effects are profound, affecting career prospects, income potential, and the ability to break free from the cycle of poverty.

The erosion of meritocracy is a subtle yet deeply insidious aspect of economic inequality. The belief that hard work and talent alone can lead to success is undermined by the stark reality of a society where wealth dictates access to opportunities.

1. Diminished Upward Mobility: In a landscape dominated by economic inequality, the ladder of social mobility is increasingly out of reach for many. The concentration of wealth not only limits access to opportunities but also consolidates power and influence within a narrow segment of society. This concentration creates a self-perpetuating elite, where opportunities for advancement, professional networks, and influential social circles are largely inaccessible to those outside the economic upper echelons.

2. Barriers to Education and Employment: The gateway to opportunity—education—becomes a fortified barrier rather than a bridge. With higher education costs soaring, the dream of pursuing advanced degrees becomes unattainable for many, locking out a significant portion of the population from the benefits of higher education, including better job prospects and higher earning potential. This barrier extends into the job market, where the best opportunities often require qualifications that are increasingly difficult to obtain without substantial financial resources.

The manifestations of economic inequality paint a picture of a society at a crossroads. The deep divisions and disparities that mark this landscape call for urgent and comprehensive measures to bridge the gap, fostering a more equitable, healthy, and cohesive society. Addressing these manifestations requires a dynamic approach, targeting the root causes of inequality and ensuring that opportunities for health, education, and economic advancement are accessible to all, irrespective of their financial standing.

Consequences for Society and Democracy

Economic inequality not only deepens social divides but also poses a formidable threat to the very foundations of democracy. The correlation between wealth accumulation and political power is stark, manifesting in the ability of the affluent to influence policy-making, elections, and the broader political discourse. This concentration of influence in the hands of a few dilutes the democratic principle of equality, where each citizen's voice and vote are supposed to carry equal weight.

1. Disproportionate Influence on Policy and Elections: Wealthy individuals and corporations often have the means to support political campaigns, lobby for favorable legislation, and shape public opinion through media ownership. This financial clout allows them to sway policy in ways that serve their interests, often at the expense of the public good. As a result, policies that could address economic disparities—such as tax reforms, wage legislation, and social welfare programs—are frequently sidelined or watered down.

2. Disenfranchisement of the General Populace: The outsized influence of the wealthy erodes trust in democratic institutions and processes. When the majority of the population perceives that their needs and voices are secondary to the interests of a wealthy minority, it undermines faith in democracy. This sense of disenfranchisement can lead to apathy and disengagement from the democratic process, weakening democracy from within.

The consequences of economic inequality extend beyond the erosion of democratic ideals, manifesting in tangible societal unrest and deepening divisions. Historical precedents show that significant economic disparities often lead to social instability.

1. The Catalyst for Social Unrest: Significant economic disparities can act as a catalyst for social unrest. When large segments of the population feel marginalized and perceive the economic system as inherently unjust, it can lead to widespread protest and civil unrest. These expressions of discontent are often a response to immediate economic pressures, such as unemployment, poverty, and lack of access to essential services, but they are underpinned by deeper issues of inequality.

2. Fueling Division and Polarization: Economic inequality does not merely separate society along economic lines; it also fosters an "us versus them" mentality that can seep into the political realm. This mentality exacerbates political polarization, making it increasingly difficult to find common ground or engage in constructive dialogue. The divisions sown by economic disparities go beyond mere disagreement, threatening the very cohesion of society and making it more challenging to address collective challenges.

The consequences of economic inequality for society and democracy are profound and far reaching. Addressing these consequences requires a concerted effort to tackle the root causes of inequality, ensuring that the economic system works for all members of society. Policies aimed at redistributing wealth, ensuring fair wages, and providing equal access to opportunities are critical. Additionally, safeguarding the integrity of democratic processes and ensuring that all voices are heard and valued are fundamental to mitigating the adverse effects of inequality. Only through such comprehensive measures can the cycle of disparity, disenfranchisement, and division be broken, paving the way for a more equitable and democratic society.

Case Studies: From Silicon Valley to the Slums of Mumbai

Silicon Valley stands as a modern epitome of both technological progress and economic disparity. Here, amidst the global headquarters of tech giants and startups alike, billionaires and tech moguls reside in sprawling estates, showcasing the lavish lifestyles that technological innovation and entrepreneurship can afford. However, this picture of abundance and prosperity is incomplete without acknowledging the stark contrast presented by the visible homelessness and pervasive poverty just a few miles from these opulent enclaves.

1. Wealth Concentration in the Tech Sector: Silicon Valley's economy, driven by the tech industry, has created immense wealth for a relatively small group of individuals. The success stories of startups turned behemoths have become legend, yet the wealth generated has largely remained within a tight-knit circle of creators, investors, and a few high-ranking executives. This concentration of wealth highlights a significant challenge in the notion that technological advancement inherently leads to broader societal prosperity.

2. Socioeconomic Disparities: The economic boom in Silicon Valley has exacerbated living costs, making it increasingly difficult for non-tech workers and even middle-class professionals to afford housing in the region. This economic stratification has led to a growing homelessness crisis, with thousands living in makeshift encampments, underscoring the failure of the region's economic growth to benefit all its residents equally.

Mumbai, India's financial capital, presents a striking case of economic inequality on a global scale. In this bustling metropolis, extreme wealth and severe poverty exist side by side, illustrating the complex nature of economic disparity in one of the world's most populous cities.

1. A Tale of Two Cities: On one end of the spectrum stands Antilia, a billion-dollar private residence towering over the city with its architectural grandeur and luxury. On the opposite end lies Dharavi, one of the largest slums in Asia, where approximately a million people live in dense, informal settlements. The proximity of such extreme living conditions within the same urban landscape starkly

highlights the chasm between the city's richest and poorest residents.

2. Economic Growth and Inequality: Mumbai's story reflects the broader narrative of India's economic ascent, marked by rapid growth and significant achievements in various sectors. However, this progress has not been evenly distributed. The city's infrastructure and services strain under the weight of its burgeoning population, with many of its residents lacking access to clean water, sanitation, and stable housing. This disparity raises critical questions about the inclusivity and sustainability of economic growth models that allow such stark inequalities to persist.

The case studies of Silicon Valley and Mumbai offer profound insights into the manifestations and consequences of economic inequality in both developed and developing contexts. They challenge us to reconsider the metrics of success and progress in an era where technological innovation and economic growth can coexist with deep-seated poverty and exclusion. Addressing the disparities highlighted by these case studies requires a concerted effort from governments, businesses, and civil societies to create more inclusive economic systems that ensure the benefits of growth and innovation are widely shared across all strata of society.

Paths Forward: Addressing the Chasm

The growing wealth chasm poses significant challenges to our society, but it also presents an opportunity to enact meaningful change. Addressing this disparity requires a versatile approach, encompassing policy reforms, educational initiatives, and corporate responsibility. Each of these elements plays a crucial role in bridging the gap, creating a foundation for a more equitable and prosperous society.

To effectively combat economic inequality, comprehensive policy interventions are essential. These policies must be designed to redistribute wealth more equitably and ensure that all members of society have the opportunity to thrive.

1. Progressive Taxation: Implementing a more progressive tax system, where the wealthiest individuals and corporations contribute a fairer share of their earnings, is a critical step towards redistributing wealth. Such measures can include higher tax rates on capital gains, luxury goods, and large inheritances, ensuring that those who benefit most from the economy contribute proportionately to its sustainability.

2. Minimum and Living Wages: Policies that raise the minimum wage to a living wage standard can have a profound impact on lifting individuals and families out of poverty. By ensuring that all workers receive a wage that meets basic living standards, we can stimulate economic growth from the bottom up, promoting a more inclusive economy.

Equitable access to education and vocational training is fundamental to leveling the playing field, allowing individuals from all backgrounds to compete and succeed in a dynamic global economy.

1. Expanding Educational Access: Scholarships, grants, and subsidized education programs can play a pivotal role in ensuring that financial barriers do not prevent talented individuals from pursuing higher education. By investing in education, we can equip more people with the skills and knowledge necessary to thrive in the 21st-century economy.

2. Vocational and Technical Training: Not everyone follows a traditional academic path, and vocational training offers a valuable alternative. By expanding access to technical education and apprenticeship programs, we can prepare individuals for high-demand jobs in industries such as technology, healthcare, and renewable energy.

Corporations have a significant role to play in mitigating economic inequality. Beyond the pursuit of profit, businesses have a responsibility to their employees, their communities, and society at large.

1. Fair Wages and Employee Benefits: Companies can demonstrate their commitment to economic equity by providing fair wages, comprehensive benefits, and opportunities for advancement to all employees. Investing in the workforce is not only a moral imperative but also a strategic business decision that can enhance productivity and loyalty.

2. Community Investment and Policy Advocacy: Corporations can leverage their resources and influence to support community development and advocate for policies that benefit society as a whole. From investing in local infrastructure and education to supporting environmental sustainability and healthcare initiatives, businesses can contribute to creating a more equitable world.

The path forward requires concerted efforts from all sectors of society—government, business, and individuals alike. By embracing progressive policies, fostering educational opportunities, and advocating for corporate responsibility, we can address the underlying causes of economic inequality. This moment in history offers a chance to redefine our values and priorities, aiming for a future marked by equity, compassion, and shared prosperity. The legacy we leave for future generations depends on the actions we take today to bridge the wealth chasm and build a more inclusive society.

Chapter 6: Connected Yet Isolated - The Technology Paradox

As dawn breaks on the digital age, society finds itself at the crossroads of unprecedented connectivity and burgeoning isolation, encapsulating a paradox that defines our time. This chapter,

"Connected Yet Isolated - The Technology Paradox," looks into the intricate relationship between technological advancements and the nuanced shades of human connection and solitude they foster.

The Genesis of Connectivity

The late 20th and early 21st centuries heralded an era that would redefine human interaction and societal structure: the advent of the internet. Emerging as a revolutionary technology, the internet was poised to connect individuals across the globe in ways previously unimaginable, transcending physical distances and cultural barriers. This period of digital enlightenment promised an unprecedented era of global unity, where information, ideas, and emotions could flow freely across the digital ether, knitting the world together in a web of shared understanding and empathy.

As the new millennium progressed, this promise seemed to materialize through the rapid proliferation of social media and communication platforms. From the early forums and chat rooms to the dynamic and multimedia-rich environments of today's social networks, these platforms have been instrumental in weaving a digital tapestry of interconnected lives. They enabled stories, personal milestones, and collective experiences to be shared and celebrated across continents, creating a global narrative that reflected the diversity and complexity of human experience in the digital age.

The early visions of a connected world envisioned by internet pioneers were coming to life, painting a picture of a global village where everyone could be a neighbor, irrespective of geographical distances. This digital landscape offered a new frontier of connectivity, where voices that were once marginalized or silenced could find an audience, and individuals seeking community could find belonging. The burgeoning networks of communication not only facilitated the exchange of information but also fostered the formation of digital communities bound by interests, beliefs, or simply the human need to connect and share.

However, this era of digital connectivity also ushered in unforeseen challenges and paradoxes. As the fabric of digital connections grew denser and more complex, questions about the nature of these connections and their impact on human relationships and societal cohesion began to surface. The very tools that promised to bring people closer together also had the potential to isolate individuals, create echo chambers, and amplify divisions. The nuanced shades of human connection, once enriched by the subtleties of face-to-face interactions, were now navigated through the filters and algorithms of digital platforms.

The genesis of connectivity, marked by the advent and proliferation of the internet and social media, thus stands as a pivotal chapter in the story of the digital age. It represents a moment of boundless optimism and potential, shadowed by the complexities and challenges of translating human connection into the digital realm. This chapter sets the stage for a deeper exploration of the technology paradox, where the lines between connectedness and isolation blur, and the quest for genuine human connection in the digital age continues to evolve.

The Illusion of Connection

As the digital age matured, the initial euphoria that surrounded the burgeoning connections across the globe began to wane, giving way to a more somber reflection on the nature and quality of these digital ties. Technology, with its dazzling promise to shrink the world and bring distant cultures and communities into the intimacy of our living rooms, succeeded in ways many had not dared to dream. Yet, this triumph of connectivity was accompanied by an unforeseen consequence: the ushering in of an era characterized by superficial interactions.

The sheer quantity of connections exploded as social networks proliferated, enveloping virtually every aspect of daily life. However, it became increasingly apparent that the depth, quality, and meaningfulness of these connections often diminished. The platforms that were designed to foster networking and build communities began to double as stages for presenting curated versions of reality. These polished, edited, and often idealized portrayals of life did more than just showcase the highlights of human experience; they distorted perceptions and created an environment ripe for comparison and discontent.

This phenomenon, sometimes referred to as the "highlight reel" effect, meant that the complex, varied nature of individual lives was reduced to a series of snapshots, carefully selected to project success, happiness, and fulfillment. The ubiquitous nature of such content on social media platforms led to an environment where the normalcy of struggle, failure, and the mundane aspects of life were underrepresented, if not entirely absent. As users engaged with these platforms, many found themselves ensnared in a cycle of comparison, measuring their own unfiltered reality against the sanitized and embellished lives of others. This cycle often resulted in feelings of inadequacy, jealousy, and loneliness, highlighting a paradoxical outcome: in a world more connected than ever before, individuals felt increasingly isolated.

The interactions facilitated by these platforms, while numerous, frequently lacked the depth and resonance of face-to-face conversations. The ease of clicking "like," sending a brief comment, or sharing emojis created a semblance of interaction that was instantaneous and effortless but often shallow. The richness of human communication, conveyed through tone of voice, facial expressions, and body language, was lost in translation to the digital medium. What remained were connections that, while convenient and wide-reaching, were stripped of the nuances and emotional depth that forge meaningful human relationships.

The era of the digital age, therefore, brought with it an illusion of connection. It created a world where people could boast hundreds, if not thousands, of "friends" or "followers" yet feel profoundly alone. As society grappled with the ramifications of this new paradigm, it became clear that the challenge was not merely technological but deeply human. The quest for genuine connection, understanding, and community in the digital age called for a reevaluation of how technology is used to mediate human relationships and a renewed emphasis on the qualities that make those connections truly enriching and meaningful.

Echo Chambers and Fragmentation

The digital revolution, while promising to connect the world in unprecedented ways, has also given rise to one of the most insidious phenomena of our time: the formation of echo chambers. These digital enclaves, forged by the very algorithms designed to personalize and enhance the user experience on social media platforms, have inadvertently narrowed the spectrum of encountered ideas and information. Instead of opening windows to the diverse panorama of human thought, these algorithms have often shut the blinds, reinforcing pre-existing beliefs and biases, and isolating users within their informational cocoons.

This inadvertent narrowing of perspectives has profound implications for public discourse and individual cognition. By filtering content to align with users' existing preferences and opinions, social media platforms create a feedback loop that amplifies sameness and mutes difference. The consequence is a digital segregation that polarizes communities and individuals, fragmenting the social fabric into isolated silos of thought. This fragmentation is not merely a division of opinions but a division of realities, where groups operate with fundamentally different sets of facts and assumptions about the world.

Moreover, the sense of community and belonging that these echo chambers might initially provide is illusory. While they offer the comfort of agreement and validation, they also foster an environment where dissenting voices are silenced or marginalized. This homogenization of thought contributes to a sense of isolation among individuals, who find themselves surrounded by a mirage of consensus that lacks the depth and challenge of genuine human interaction. The absence of divergent viewpoints not only impoverishes personal growth and understanding but also undermines the democratic processes that rely on the robust exchange of ideas.

Echo chambers thus represent a paradoxical outcome of the digital age: in seeking to connect us more closely to our interests and beliefs, technology has, in some cases, isolated us from the broader spectrum of human experience and from each other. This fragmentation of the digital landscape into echo chambers of agreement and sameness poses significant challenges to societal cohesion, understanding, and empathy. As the digital age matures, addressing the issue of digital segregation and fostering environments that encourage diversity of thought and genuine engagement with differing viewpoints becomes not just a technological challenge but a societal imperative.

The Psychological Landscape

The paradox that technology, while promising unprecedented levels of connection, paradoxically breeds isolation, is most starkly evident in the realm of mental health. Recent studies have drawn a direct line between heavy social media usage and an escalation in feelings of loneliness, anxiety, and depression, with these effects being particularly pronounced among younger generations. These individuals, who have grown up immersed in digital culture and navigate these spaces with unparalleled fervor, find themselves at the epicenter of a mental health crisis exacerbated by the very tools designed to bring them closer to others.

This crisis stems from the relentless exposure to idealized representations of life on social media platforms. Users are inundated with images and narratives that depict perfect lifestyles, bodies, and success stories, creating a skewed reality that can feel both unattainable and alienating. The pressure to participate in this digital masquerade, to curate one's online presence meticulously, contributes to a pervasive sense of inadequacy. Many users find themselves performing happiness and success in an attempt to match the curated lives they see online, all the while feeling increasingly disconnected from the authenticity of their own experiences.

The impact of this digital disillusionment on mental health cannot be understated. The constant comparison to others' highlight reels on social media platforms can erode self-esteem and exacerbate feelings of isolation, even in the midst of a seemingly connected world. The irony is profound: in an age where expressing oneself and reaching out for connection can be done with the click of a button, genuine human connections can feel more elusive than ever. The digital landscape, with its promise of community and shared experience, often morphs into a theater of comparison and competition, leaving individuals feeling more alone in their struggles and less connected to the authentic experiences of those around them.

Addressing the psychological landscape shaped by the digital age requires a diverse approach. It calls for increased awareness of the impact of social media on mental health, the cultivation of digital literacy that empowers users to navigate online spaces with a critical eye, and the fostering of environments, both online and offline, that prioritize genuine human connection and the sharing of authentic experiences. As society continues to grapple with the complexities of the digital age, the need for strategies that support mental health and foster real connections becomes increasingly paramount, highlighting the necessity of bridging the gap between the promise of technology and the reality of its impact on our psychological well-being.

Searching for Authenticity

Amidst the swirling currents of the digital age, where the paradox of connected isolation looms large, a counter-movement has begun to take shape. This movement, emerging from the depths of the crisis of connection, advocates for a reevaluation of our relationship with technology, calling for a more mindful and authentic engagement with the digital world. It recognizes the profound impact that superficial connections and curated online personas have on our mental health and societal wellbeing and proposes a path toward more meaningful digital and real-world interactions.

Initiatives aimed at fostering real-world interactions have gained momentum, encouraging people to step away from their screens and engage in face-to-face conversations, community activities, and nature immersion. These initiatives underscore the irreplaceable value of direct human contact, which conveys nuances and emotions that digital communication often fails to capture.

Simultaneously, there has been a conscious effort to reshape digital spaces to prioritize depth over breadth in connections. New platforms and features are being designed with the intention of facilitating more genuine interactions, moving away from the metrics of likes and followers to focus on the quality of connections and conversations. These digital environments aim to create spaces where users can share and engage with each other's experiences in a more authentic and empathetic manner.

The rise of digital detoxes and mindfulness apps is a testament to the growing awareness of the need to balance our online lives with our innate human need for genuine, face-to-face interactions. Digital detoxes, which encourage taking breaks from electronic devices, help individuals reconnect with themselves and their immediate physical surroundings, providing a respite from the constant digital chatter. Mindfulness apps, on the other hand, offer tools for managing digital consumption and enhancing mental wellbeing, promoting practices that help ground users in the present moment.

This search for authenticity in the digital age is not a rejection of technology but a call to use it in a way that enhances, rather than diminishes, our human experience. It acknowledges the incredible potential of digital tools to connect us across vast distances, while also reminding us of the value of deep, meaningful interactions that nourish our psychological and emotional wellbeing. As this movement grows, it paves the way for a future where technology serves to enrich our lives and connections, bridging the gap between the digital and the authentic in a harmonious balance.

Reimagining Digital Spaces

As we navigate the complexities of the digital age, the quest for a future where technology enhances rather than detracts from human interaction becomes increasingly paramount. This vision requires a reimagining of digital spaces, where the design and function of technology prioritize the richness of human experience and the depth of our connections. It's about shifting the paradigm from one that values the quantity of connections to one that cherishes their quality, fostering environments that encourage meaningful exchanges and sustained relationships.

Innovations in virtual and augmented reality (VR and AR) are at the forefront of this transformation, offering tantalizing glimpses of how technology can bridge the chasm of physical distance without sacrificing the intimacy and authenticity of human connection. These technologies have the potential to create immersive experiences that replicate the nuances of face-to-face interactions, from the subtlety of body language to the warmth of shared moments in a shared space. By enhancing our sensory engagement with digital environments, VR and AR could redefine the essence of presence, making distant interactions feel tangibly close and deeply personal.

Beyond VR and AR, the reimagining of digital spaces involves the development of platforms and tools that actively facilitate deep, sustained connections. This could mean algorithms that prioritize content fostering genuine conversation and understanding, or social networks designed to support smaller, more intimate communities where meaningful interactions can flourish. It's about creating digital environments that reflect and respect the complexity of human relationships, spaces where authenticity is valued over artifice, and where individuals are encouraged to present themselves as they are, rather than as idealized versions for consumption.

The challenge of this reimagining is not insignificant, requiring concerted efforts from technologists, designers, sociologists, and policymakers. It calls for a holistic approach that considers the psychological, social, and ethical implications of how technology is integrated into our lives. By placing human connection at the heart of technological innovation, we can create digital spaces that not only connect us across distances but also deepen the bonds that unite us, ensuring that technology serves as a bridge to more meaningful relationships and a more connected world.

Policy, Education, and Beyond

The journey through the landscapes of connection and isolation in the digital age brings us to a critical juncture: the recognition that navigating the technology paradox requires actions that extend far beyond the individual. It demands a collaborative effort that encompasses policy interventions, educational reform, and a societal shift towards a more mindful engagement with technology.

Policy interventions play an indispensable role in mitigating the excesses and unintended consequences of digital connectivity. This includes implementing regulations that ensure social media platforms operate transparently and ethically, prioritizing user well-being over engagement metrics and advertising revenue. Protecting user data from exploitation and ensuring privacy are also paramount, as these factors significantly impact the quality of online interactions and the sense of safety in digital spaces. Moreover, policies that encourage the development and deployment of technology in ways that foster genuine human connection can serve as a catalyst for positive change.

Equally crucial is the role of education in preparing individuals to navigate the complexities of digital life. Integrating digital literacy and ethics into educational curricula from an early age is essential for developing a generation that is not only technologically proficient but also aware of the social, psychological, and ethical dimensions of digital interactions. This education should aim to cultivate empathy, critical thinking, and a nuanced understanding of how technology impacts individual lives and society at large.

As "Connected Yet Isolated - The Technology Paradox" concludes, it extends a call to action to individuals, communities, and societies to reclaim the transformative promise of technology as a tool for genuine connection. This involves recognizing the pitfalls of our current engagement with digital spaces and actively seeking out solutions that emphasize real-world interactions and authentic relationships. It's about creating an environment where technology serves as a bridge to understanding and unity, rather than a barrier to it.

By embracing this call to action, we can begin to bridge the chasm between connectivity and isolation. The path forward involves leveraging the true potential of technology to unite us—fostering a world where digital spaces complement and enhance, rather than substitute, the richness of human experience. In doing so, we not only navigate the technology paradox but also harness its vast potential to create a more connected, empathetic, and understanding world.

Chapter 7 : Mind Traps - The Mental Health Crisis and Social Media

In the digital panorama of the 21st century, the rise of social media platforms has transformed human interaction, forging new pathways for communication, creativity, and community. Yet, beneath this digital veneer lies a less visible, more insidious phenomenon: a burgeoning mental health crisis that threads through the fabric of societies worldwide. As mental health issues escalate, particularly among the youth, the omnipresent influence of social media platforms—designed to captivate and engage—casts a long shadow over individual well-being. Historians of the future, armed with hindsight, may not look back on this era fondly. They will scrutinize our collective and individual responses to this dual-edged sword: a tool that connects yet isolates, empowers yet diminishes. This chapter looks into this complex narrative, exploring the mental health crisis in the age of social media, its impacts, societal responses, and the lessons that may guide future generations toward a more balanced coexistence with digital technology.

Historical Context

The digital revolution marks a pivotal chapter in the annals of human history, characterized by the rapid adoption of the internet and personal computing devices. This era unfolded a new dimension of connectivity, laying the groundwork for what would become the social media landscape. The late 20th and early 21st centuries witnessed the birth and meteoric rise of platforms like Facebook, Twitter, and Instagram. Initially hailed as groundbreaking avenues for communication, creativity, and self-expression, these platforms swiftly embedded themselves into the fabric of daily life, altering the ways in which individuals connect, share, and perceive the world around them.

As the digital age progressed, the evolution of social media platforms revealed a more complex narrative regarding their influence on society. What began as digital spaces for enhancing personal connections and broadening one's social horizon gradually morphed into arenas of constant comparison, relentless pursuit of validation, and an ever-present cycle of information and misinformation. This transformation raised significant concerns about the psychological and emotional well-being of users, particularly as instances of cyberbullying, social isolation, and the pressure to conform to unrealistic standards of beauty and success proliferated.

The historical trajectory of social media, from its nascent stages of optimistic potential to its current state of contentious impact, underscores a nuanced dynamic. These platforms, while continuing to offer unprecedented opportunities for engagement and knowledge sharing, also serve as catalysts for a deeper examination of their long-term effects on mental health and societal well-being. As we navigate the digital epoch, the historical context of social media's rise serves as a crucial lens through which we can evaluate its dual-edged legacy: a testament to human ingenuity and a mirror reflecting the complex challenges of the digital age.

Psychological Impact

The magnetic pull of social media is intricately tied to its provision of instant gratification. Likes, shares, comments, and followers serve as digital affirmations of social acceptance and personal worth, engaging users in a perpetual quest for validation. This quest, while seemingly benign in its digital context, activates dopamine-driven feedback loops similar to those observed in other forms of addiction. Such mechanisms make social media not just a tool for connection but a potential catalyst for compulsive behavior, drawing users back repeatedly with the promise of immediate rewards.

Beyond the addictive potential of these platforms, there lies a more insidious effect tied to the very nature of the content shared. Social media feeds are often curated showcases of success, beauty, and apparent happiness, presenting an idealized glimpse into the lives of others. This curated reality sets a benchmark for personal achievement and lifestyle that is often unattainable or misleading. The constant exposure to such highlights has been linked to a pervasive culture of social comparison, where users measure their own worth and success against the polished personas and achievements of their peers.

This comparison does not occur in a vacuum. It seeds and nurtures feelings of inadequacy, envy, and dissatisfaction, contributing to a wider spectrum of psychological effects. Anxiety and depression, in particular, have been noted as significant outcomes of this relentless comparative mindset, with users finding themselves trapped in a cycle of self-doubt and negative self-perception. The irony of social media is thus its ability to connect us more broadly than ever before while simultaneously fostering a sense of isolation and inadequacy within the very networks it creates.

The psychological impact of social media is a complex interplay of addiction, comparison, and the pursuit of unattainable ideals. As these digital platforms continue to evolve, so too does the need for a deeper understanding of their long-term effects on mental health and well-being. Recognizing and addressing the negative psychological outcomes of social media use is crucial in navigating the balance between its benefits and the challenges it poses to individual and societal health.

The Paradox of Connection

At the heart of social media's allure is the promise of enhanced connectivity—a digital bridge linking individuals across the globe, fostering a sense of community and shared experience. Yet, this promise bears an ironic twist: rather than mitigating the human experience of loneliness, social media often exacerbates it, ushering in a new age of isolation concealed beneath the veneer of constant communication.

The interactions that occur within these digital arenas, though vast in number, frequently lack the depth and emotional resonance characteristic of face-to-face connections. Comments, likes, and shares, while signaling attention and recognition, fall short of conveying genuine empathy, understanding, and emotional support.

The result is a peculiar form of social fulfillment, one that is broad yet shallow, leaving individuals feeling disconnected and isolated despite the seemingly ever-present network of digital contacts.

This paradoxical effect stems from the inherent limitations of digital communication. The nuanced cues of human interaction—tone of voice, body language, facial expressions—are diminished or altogether absent in online exchanges. Without these cues, the richness of human communication is lost, replaced by a simulacrum that fails to satisfy our intrinsic need for social bonding and emotional connection.

The curated nature of social media content contributes to this sense of isolation. As users present idealized versions of their lives, the platform becomes a stage for performance rather than a space for authentic connection. The resulting environment, where reality is glossed over and vulnerabilities are hidden, further impedes the formation of meaningful connections, leaving individuals feeling lonely even in a crowd of digital onlookers.

The paradox of connection, thus, highlights a critical challenge of the digital age: reconciling the desire for widespread connectivity with the need for genuine, emotionally resonant relationships. As we navigate this digital landscape, understanding and addressing the nuances of this paradox will be essential in fostering a healthier, more connected society.

Prevalence of Mental Health Issues

In the contemporary landscape of global health, mental health disorders have emerged as a silent epidemic, casting long shadows across societies worldwide. This trend is particularly pronounced among adolescents and young adults, a demographic that finds itself at the epicenter of a burgeoning mental health crisis. Health organizations across the globe have sounded alarms over the escalating incidence rates of depression, anxiety, and a spectrum of other psychological disorders, painting a concerning picture of the state of mental well-being in the 21st century.

This uptick in mental health issues is not solely attributable to advancements in diagnostic practices or increased awareness among the public and healthcare professionals. While such factors have indeed played a role in identifying and acknowledging mental health disorders, the data suggests a genuine and widespread increase in the prevalence of these conditions. The implications are profound, signaling a societal shift that warrants close attention and immediate action.

A closer examination of the data reveals a stark reality: a significant portion of the global population is currently battling mental health challenges, with the scales tipping alarmingly towards younger demographics. Adolescents and young adults, navigating the complexities of modern life, digital integration, and social pressures, are finding themselves disproportionately afflicted by mental health disorders. This demographic, ideally at the peak of their physical health and embarking on the journey of shaping their futures, is instead confronted with the debilitating effects of mental health issues, affecting their education, relationships, and overall quality of life.

The reasons behind this alarming trend are multidimensional, ranging from the pressures of contemporary societal expectations to the pervasive influence of digital media and social networking sites. The cumulative effect of these factors, coupled with the challenges inherent in transitioning from adolescence to adulthood, creates a perfect storm for the onset and exacerbation of mental health disorders.

As we grapple with the implications of this rising trend, it becomes imperative to view the prevalence of mental health issues not just as a health crisis but as a complex societal challenge that requires a multi-faceted approach to address. From enhancing access to mental health services to fostering supportive environments that reduce stigma and encourage open discussions about mental well-being, the path forward demands concerted efforts from individuals, communities, and policymakers alike.

Linkages to Social Media

The intertwining of social media use with mental health issues is a phenomenon that is gaining substantial support from the scientific community. A growing body of research underscores the correlation between heavy social media engagement and the exacerbation of mental health disorders, particularly depression and anxiety. This connection, far from being coincidental, reveals the profound impact that digital environments can have on individual psychological well-being.

One of the primary mechanisms through which social media affects mental health is the phenomenon known as the fear of missing out (FOMO). This anxiety, driven by the perception that others are experiencing more fulfilling lives, can lead to persistent feelings of inadequacy and dissatisfaction. Social media platforms, with their endless streams of curated content showcasing highlights of others' lives, are fertile ground for FOMO, exacerbating the pressure to measure up to an often unattainable standard of success and happiness.

Additionally, online harassment and cyberbullying present a direct link to depressive symptoms among users. The anonymity and distance afforded by digital platforms can embolden negative behavior, subjecting individuals to criticism, harassment, and abuse that they might not encounter in face-to-face interactions. The impact of these experiences can be deeply traumatic, leading to long-term psychological distress.

Furthermore, the curated nature of social media content contributes to distorted self-perceptions and dissatisfaction. The selective presentation of life's highlights creates a skewed reality, where everyday struggles and failures are obscured, leading users to question their own life experiences and worth. This distortion plays a significant role in the development of low self-esteem and body image issues, further implicating social media in the broader mental health crisis.

The linkages between social media use and mental health are complex and nuanced. The digital age brings with it unprecedented challenges to mental well-being, driven by the unique pressures and dynamics of online interaction. Recognizing and addressing these linkages is crucial in mitigating the adverse effects of social media on mental health, paving the way for healthier digital environments.

Government and Policy Responses

In the face of a mounting mental health crisis exacerbated by social media, governments and policymakers worldwide have initiated varied responses to mitigate its impacts. These efforts have largely focused on regulatory measures designed to protect online users, particularly vulnerable groups such as adolescents and young adults. Among these measures are laws and policies aimed at combating cyberbullying, enforcing age restrictions on social media use, and ensuring that platforms take active steps to protect their users' well-being.

Some countries have gone further, introducing innovative policies that include mandatory downtime for youth on social media platforms, mechanisms for reporting and removing harmful content, and requirements for social media companies to demonstrate their platforms do not harm children's mental health. Additionally, there have been initiatives to promote digital literacy, helping users navigate social media in healthier ways and understand the potential psychological impacts of their online behaviors.

Despite these efforts, the effectiveness of governmental and policy interventions in curbing the mental health issues linked to social media use remains under scrutiny. Critics point out that many of these measures, while well-intentioned, only scratch the surface of a deeply rooted problem. They argue that legislative actions often fail to tackle the underlying causes of the crisis, such as societal pressures magnified by social media, the inherent design of platforms that encourage addictive use, and the broader cultural normalization of sharing and comparing personal lives online.

Furthermore, there is a call for more comprehensive strategies that not only regulate but also educate and provide support. This includes integrating mental health education into school curriculums, offering resources for mental health support tailored to the digital age, and fostering a societal shift towards more authentic and supportive online environments. The challenge for governments and policymakers, then, is not just to regulate but to envision and implement holistic approaches that address the complex nature of the mental health crisis in the digital age.

Social Media Companies: Navigating Ethical and Business Dilemmas

As central actors in the digital landscape, social media companies find themselves at the intersection of public scrutiny and ethical responsibility regarding the mental health crisis. Acknowledging the growing concerns, some platforms have initiated steps to mitigate negative impacts on user well-being. These measures include introducing content control options that allow users to filter what they see, tools for reporting harmful behavior, and features designed to encourage breaks from prolonged use. While these initiatives represent a shift towards more conscientious platform management, they are frequently criticized for being more cosmetic than curative, addressing symptoms rather than the root causes of the crisis.

The crux of the challenge lies in the inherent conflict of interest between profit-driven motives and the imperative to protect and enhance user well-being. Social media companies operate in a highly competitive market where user engagement directly correlates with advertising revenue. This economic model incentivizes the design of algorithms that prioritize content likely to keep users online longer, often at the expense of their mental health. The ethical considerations of such algorithmic curation practices have sparked intense debate, highlighting a contentious issue: Can social media platforms reconcile their business objectives with the responsibility to safeguard user mental health?

This question becomes even more complex when considering the opaque nature of algorithmic decision-making, which often lacks transparency and accountability. Critics argue that without a fundamental reevaluation of how content is curated and presented, efforts to improve user well-being may remain superficial. This includes rethinking the metrics of success, moving beyond engagement and virality to incorporate indicators of positive social impact and user health.

In response, there is a growing call for social media companies to adopt a more ethically grounded approach to platform design and management. This involves not only implementing measures to protect users from harm but also actively promoting healthier patterns of use. It suggests a paradigm shift towards platforms that are not only engaging but also nurturing, supporting users' mental health rather than undermining it.

The road ahead for social media companies is fraught with complex ethical, business, and technological challenges. Navigating this terrain requires a delicate balance, a willingness to innovate, and a commitment to placing user well-being at the heart of platform design and operation. As the digital landscape continues to evolve, the actions and policies adopted by these companies will play a pivotal role in shaping the future of our online social ecosystem.

Community Support Systems: Bridging the Gap

Amidst the ongoing mental health crisis, exacerbated by the complexities of social media use, community support systems have risen as beacons of hope and resilience. In response to the gaps left by institutional and policy interventions, grassroots initiatives, online support groups, and mental health advocacy organizations have carved out spaces of support, understanding, and action. These community-led efforts highlight the potent force of collective action and the invaluable role of shared experiences in the fight against the isolating impacts of mental health challenges and the digital world.

Grassroots initiatives often emerge from personal experiences and the recognition of unmet needs within communities. They offer localized solutions and support, tailored to the unique contexts and challenges faced by their members. From local meetups and workshops to community mental health days, these initiatives foster environments where individuals feel seen, heard, and understood, away from the often impersonal nature of institutional responses.

Online support groups have similarly become critical resources, leveraging the very platforms that contribute to the crisis for positive ends. These groups provide safe havens for individuals to share their experiences, challenges, and victories over mental health struggles, free from the judgments and pressures of broader social media landscapes. Through shared stories and resources, participants find solace and strength, learning that they are not alone in their journeys.

Mental health advocacy organizations play a pivotal role in amplifying the voices of those affected by the crisis. Through campaigns, research, and policy advocacy, they work to shift public discourse, influence policy, and secure resources for mental health support. These organizations bridge the gap between individuals and systemic change, advocating for a society where mental health is prioritized and stigma is eradicated.

The essence of community support systems lies in their emphasis on solidarity, shared experiences, and collective well-being. They remind us that, in the face of challenges both old and new, there is immense power in coming together, sharing our stories, and supporting one another. As we navigate the complexities of the digital age and its impact on mental health, these community networks serve as crucial pillars of strength, offering hope and pathways to resilience in the midst of societal and technological change.

Building Resilience in the Digital Age

As we grapple with the mental health crisis in the shadow of the digital revolution, the cultivation of individual resilience emerges as a paramount endeavor. Resilience—the capacity to recover from difficulties, adapt to change, and maintain or regain mental well-being in the face of adversity—is not merely an innate trait but a skill set that can be developed and strengthened over time. Within the context of the pervasive influence of social media and its associated challenges, building resilience is both a protective and proactive strategy. It involves a dynamic approach, centered around digital literacy, mindful engagement with technology, and the preservation of a balanced life that honors both online and offline experiences.

Promoting Digital Literacy: Understanding the mechanisms, motivations, and manipulations inherent in social media platforms is the first step toward resilience. Digital literacy empowers individuals to navigate the online world with a critical eye, recognizing the difference between constructive and destructive digital engagement. It includes an awareness of how algorithms work, the psychological impacts of social media use, and the importance of questioning the authenticity and intention behind digital content.

Encouraging Critical Engagement: Encouraging a more mindful and critical approach to social media use goes hand in hand with digital literacy. This means fostering an environment where individuals feel empowered to question the value and impact of their online interactions. Are these interactions enriching their lives, or contributing to stress and anxiety? Critical engagement also involves recognizing and resisting the pressures to conform to unrealistic standards or participate in negative online behaviors.

Advocating for a Balanced Online-Offline Life: Perhaps one of the most crucial strategies for building resilience is advocating for and practicing a balanced life that values both online and offline experiences. This balance is vital for mental well-being, encouraging individuals to engage in meaningful offline activities that foster real-world connections, hobbies, and relaxation. Setting boundaries around social media use—such as designated tech-free times or spaces—can help reinforce this balance, making room for face-to-face interactions and activities that contribute to physical and mental health.

Incorporating Mindfulness Practices and Meaningful Connections: Mindfulness practices, such as meditation, deep breathing, and yoga, can enhance emotional regulation and self-awareness, helping individuals to stay grounded in the present moment rather than getting lost in the digital world. Likewise, seeking out and nurturing meaningful offline relationships provides a sense of belonging and support that social media cannot replicate. These relationships are the cornerstone of resilience, offering emotional support, empathy, and a sense of community that bolsters mental health.

Building resilience in the digital age is an ongoing journey, one that requires awareness, intention, and the willingness to adapt and grow. By embracing these strategies, individuals can forge a path through the complexities of the digital landscape, safeguarding their mental well-being and thriving in both the virtual and real worlds.

Critical Analysis: Reflecting on the Mental Health and Social Media Crisis

As we navigate through the complexities of the 21st century, marked by the unprecedented rise of social media and a parallel increase in mental health issues, future historians may undertake a critical examination of this era. Their analysis might focus on the societal, institutional, and individual responses to the intertwining crises of mental health and digital consumption. In particular, they may pinpoint the reluctance to address the root causes of these challenges and the sluggish pace of implementing meaningful reforms as pivotal factors that exacerbated the situation.

The critical lens through which future analysts might view this period could highlight a series of missed opportunities for early intervention and systemic change. The early signs of the mental health crisis, intensified by the digital revolution, offered numerous occasions for proactive measures. Yet, a comprehensive and coordinated response was often lacking. This delay in action might be attributed to a variety of factors, including underestimation of the crisis's severity, the novelty and complexity of digital-age challenges, and conflicting interests within the spheres of technology, economy, and public health.

Future critiques may underscore the systemic failures that allowed the crisis to deepen. These include inadequate mental health resources, lack of universal access to care, insufficient regulation of social media platforms, and the slow adaptation of educational systems to incorporate digital literacy and resilience-building. The analysis could also explore how these systemic shortcomings were compounded by a broader cultural reluctance to confront uncomfortable truths about social media's role in shaping public and private life.

The interplay between social media companies' profit-driven motives and the public interest might be another focal point of criticism. The ethical dilemmas posed by algorithmic content curation, data privacy concerns, and the manipulation of user behavior for commercial gain could be viewed as significant contributors to the crisis. Historians might argue that the failure to adequately regulate and hold these companies accountable represents a missed opportunity to mitigate the negative impacts on mental health.

In retrospect, the current era's handling of the mental health and social media crisis may be seen as a cautionary tale of the consequences of inaction, misprioritization, and the failure to adapt to new realities. Future analyses might stress the importance of learning from these oversights, advocating for a more holistic approach to tackling such complex challenges. This would involve not only technological and regulatory innovations but also a cultural shift towards prioritizing mental health, fostering resilience, and reimagining our relationship with digital technologies. The hope would be that such reflections guide future generations in creating a more balanced, healthy, and equitable digital society.

Lessons for the Future: Charting a Path Toward a Healthier Digital World

As we traverse through the tumultuous landscape of the mental health and social media crisis, the lessons gleaned from this era hold profound implications for future generations. These lessons, born out of both the challenges and triumphs of our time, illuminate a path forward that prioritizes human well-being in the digital age. Reflecting on this period, there are several key insights that future societies can draw upon to foster a more harmonious relationship with digital technologies.

The Crucial Role of Digital Literacy Education: One of the fundamental lessons is the importance of embedding digital literacy education early and comprehensively within societal frameworks. Empowering individuals with the knowledge to critically navigate the digital world, understand the implications of their online activities, and recognize the psychological effects of social media use is paramount. This education should not only focus on the mechanics of digital platforms but also on fostering a healthy digital mindset, enabling users to engage with technology in ways that support their mental well-being.

Strengthening Mental Health Support Systems: The escalation of mental health issues in the context of social media underscores the urgent need for robust mental health support systems. These systems must be accessible, inclusive, and equipped to address the nuances of digital-age stressors. Investing in mental health services, research, and public awareness campaigns can build a resilient foundation, ready to support individuals through the challenges of the digital world.

Ethical Responsibilities of Technology Companies: The role of technology companies in shaping the digital landscape cannot be overstated. These entities must acknowledge their ethical responsibilities in designing and managing social media platforms. Prioritizing human well-being over engagement metrics and profit margins requires a paradigm shift—a reevaluation of success measures that includes the mental and emotional health of users. Implementing design choices that promote positive interactions and protect users from harm is a critical step toward ethical digital stewardship.

A Forward-Looking Perspective: Reflecting on the current era's shortcomings and achievements offers invaluable insights for future digital environments. By critically analyzing the impact of social media on mental health, society can aspire to develop technologies that enhance human connection without compromising mental well-being. This entails a collaborative effort among policymakers, technology companies, mental health professionals, and the community at large.

The intersection of social media and mental health presents a complex challenge, but also an opportunity for transformative change. As we look toward the future, the actions and decisions of today will undoubtedly influence the digital legacy we leave behind. By learning from the past and present, future generations can cultivate a digital landscape where technology serves as a tool for positive growth and well-being, ensuring a balanced and healthy digital existence for all.

Chapter 8: Identity and Division - The Politics of Us vs. Them

In an era marked by unprecedented connectivity and access to information, society finds itself more divided than ever along lines of identity, ideology, and belief. The politics of us vs. them has not only shaped our social landscapes but has also entrenched divisions, creating fissures that run deep through the heart of communities worldwide. This chapter seeks to unravel the nature of this complex issue, exploring the roots of identity politics, the role of media in amplifying divisions, and the profound impact on societal cohesion and governance. Through a meticulous examination, we aim to shed light on the pathways that might lead us from division towards reconciliation, envisioning a future where unity and empathy prevail over segregation and animosity.

The evolution of identity politics is a nuanced journey that reflects the changing dynamics of societal structures, beliefs, and the quest for equality and recognition. Originally springing from the fertile ground of civil rights movements in the 20th century, identity politics aimed to spotlight and rectify the systemic injustices faced by marginalized groups. These movements, pivotal in shaping modern discourse on race, gender, and sexuality, emphasized the importance of recognizing and valuing the unique experiences and challenges of these groups, advocating for equal rights and opportunities.

As the decades progressed, the landscape of identity politics underwent significant transformation. This evolution was marked by an increasing focus on the distinctiveness of individual and group identities. While initially, this served as a powerful means to highlight overlooked or suppressed voices, over time, it has also led to a more polarized environment. The emphasis on difference rather than commonality has, in some instances, fostered a perception of zero-sum dynamics within the political and social arenas. In this viewpoint, the gains of one group are often seen as coming at the direct expense of another, leading to competition rather than collaboration among communities seeking equity and recognition.

This shift towards a more divisive application of identity politics raises complex questions about the balance between celebrating diversity and achieving broader societal unity. The challenge lies in navigating these waters without negating the individual experiences of marginalized communities, while also fostering a sense of common purpose and mutual respect across diverse societal segments. As identity politics continues to evolve, its impact on social cohesion, governance, and the pursuit of equity remains a critical area of examination and dialogue.

The psychological underpinnings of the "us vs. them" mentality are both fascinating and complex, touching upon the very core of human nature and social behavior. This dichotomy, deeply embedded within our psyche, originates from an evolutionary imperative: the need to form cohesive groups that could effectively compete for resources and ensure survival. In the primitive context, identifying with an "us" provided safety, support, and strength, while distinguishing a "them" helped in recognizing potential threats and competitors.

This instinctual process of categorization extends beyond mere survival, influencing our perceptions of identity, belonging, and loyalty. It taps into basic human emotions such as fear, pride, and the need for connection, driving us to seek out similarities and shun differences. In the serene landscapes of ancient times, this might have been a straightforward mechanism for enhancing group cohesion and delineating friend from foe. However, the sophisticated structure of modern, pluralistic societies presents a far more complex scenario where these instinctual drives can lead to negative outcomes.

Xenophobia, intolerance, and conflict often find their roots in this primal division, exacerbated by perceptions of scarcity and competition. The modern world, with its constant barrage of information and interaction, highlights differences and stokes fears over economic resources, social status, and cultural dominance. This perceived scarcity, whether real or imagined, fuels competition and hostility between groups, driving a wedge deeper into the societal fabric.

Understanding these psychological underpinnings is crucial for addressing the divisive nature of contemporary identity politics and social conflicts. Recognizing that these instincts, while deeply ingrained, can be mitigated through conscious effort, education, and inclusive policies is a step toward overcoming the destructive patterns they engender. By fostering environments that emphasize shared values, mutual respect, and collective well-being, societies can begin to bridge the divides, turning the instinctual "us vs. them" into a more inclusive "we."

The "us vs. them" mentality is a pervasive aspect of human psychology, deeply entrenched in our evolutionary history. This instinctive tendency to classify individuals as either members of an in-group (us) or an out-group (them) plays a critical role in shaping our social landscapes. Historically, this mechanism served as a survival strategy, helping to foster group cohesion and identity by distinguishing allies from adversaries. However, in the complex web of modern identity politics, this binary way of thinking has taken on a new dimension, often leading to heightened divisions within society.

In the realm of identity politics, the emphasis often shifts towards highlighting differences rather than seeking common ground. This focus on distinction rather than similarity exacerbates the natural inclination towards "us vs. them" thinking, creating an environment where compromise and dialogue are not just undervalued but may be viewed with suspicion or outright hostility. The narrative that one's own group's advancement or perspectives can only be achieved at the expense of another group further fuels this division, making reconciliation and mutual understanding increasingly challenging.

The emotional bond and sense of belonging that come with group identification can amplify these divisions. Loyalty to the group can become paramount, overshadowing individual critical thinking and empathy for those outside the group. This loyalty often means that any concession or attempt to understand the "other" is seen not as a step towards bridge-building but as a betrayal of one's own group. This dynamic can lead to a vicious cycle where dialogue is stifled, and empathy is replaced with animosity, further entrenching the divide.

Understanding the roots and mechanisms of the "us vs. them" mentality is crucial for addressing the challenges it poses to social cohesion. By recognizing the power of narratives and the emotional pull of group loyalty, there is potential for creating strategies that foster empathy, highlight commonalities, and encourage a more inclusive approach to addressing societal issues. This shift requires concerted efforts in education, media representation, and leadership to move beyond binary thinking towards a more nuanced and empathetic understanding of identity and difference.

The intersection of media, technology, and societal division is a critical area of study in understanding the dynamics of modern identity politics and the "us vs. them" mentality. Digital media and technology, with their ubiquitous presence and influential power, play a significant role in both reflecting and shaping societal divides. The algorithms that underpin many of these platforms are engineered to maximize user engagement, often through the promotion of content that triggers strong emotional responses. This can lead to a reinforcement of existing beliefs and an escalation in ideological polarization, as individuals are increasingly exposed to information that aligns with their viewpoints, while opposing perspectives are filtered out.

Social media platforms, in particular, have transformed into battlegrounds for identity politics, where simplified and divisive narratives gain traction over more balanced and nuanced discourse. These platforms, by design, encourage quick, reactive responses rather than thoughtful consideration, further entrenching division. The ease of sharing information, combined with the viral nature of emotionally charged content, means that narratives emphasizing difference and conflict can spread rapidly across networks, reaching wide audiences with minimal effort.

The anonymity and physical distance inherent in digital interactions contribute to the dehumanization of those with differing opinions. Online, individuals are more likely to engage in aggressive behavior or dismissive rhetoric that they might avoid in face-to-face interactions. This detachment not only exacerbates divisions but also complicates efforts towards understanding and reconciliation. The perceived safety of a screen allows for the expression of extreme views without the immediate social repercussions that would typically moderate such interactions in real life.

Understanding the role of media and technology in sustaining societal divisions is crucial for devising strategies aimed at bridging gaps and fostering a more inclusive discourse. By acknowledging the ways in which digital platforms can amplify division, stakeholders can work towards creating digital environments that promote empathy, understanding, and nuanced discussion, countering the trend towards polarization and conflict.

Political Movements and Campaigns

The interplay between identity politics and political movements and campaigns has become increasingly salient in contemporary discourse. This section looks into the pivotal role of identity politics in shaping political landscapes, using both contemporary and historical examples to illustrate its profound impact. Identity politics, defined by the mobilization around particular aspects of group identity such as race, gender, and nationality, serves as both a powerful tool for advocacy and a potential source of division.

The Brexit referendum in the United Kingdom starkly exemplifies the divisive potential of identity politics. The campaign leading up to the referendum was heavily laden with rhetoric emphasizing national identity and sovereignty. Proponents of Brexit appealed to a sense of British identity, suggesting that leaving the European Union would allow the UK to reclaim control over its laws and borders, thereby restoring a sense of national sovereignty. This appeal to national identity successfully mobilized a significant portion of the electorate but also highlighted and intensified divisions within the country. The aftermath of the referendum has seen a Britain grappling with its identity, both internally and as part of the broader European and global communities.

Similarly, the political landscape of the United States provides ample evidence of identity politics at play. Political strategies frequently focus on mobilizing specific identity groups, leveraging shared experiences or concerns to galvanize support. For example, campaigns often target messages to racial, ethnic, or gender groups, hoping to capitalize on shared identities to secure votes. While this can empower marginalized groups and bring attention to specific issues, it also risks alienating broader coalitions and fostering divisions. The emphasis on identity can sometimes overshadow commonalities, leading to fragmented societies and polarized political spheres.

These case studies underscore the dual nature of identity politics. On one hand, it has the power to mobilize marginalized groups, giving voice to the voiceless and pushing for meaningful change. For instance, movements advocating for civil rights, gender equality, and LGBTQ+ rights have all utilized identity politics effectively to advocate for their causes. These movements have led to significant societal and legislative changes, demonstrating the positive potential of rallying around shared identities.

On the other hand, identity politics can deepen societal rifts, creating an "us versus them" mentality that fractures communities. The mobilization around specific identities can lead to exclusion, marginalizing those who do not fit within the defined identity parameters. Furthermore, it can detract from the pursuit of broader societal goals, focusing attention on differences rather than commonalities.

The influence of identity politics in political movements and campaigns highlights the complex nature of identity in shaping political discourse. While it can be a powerful force for advocacy and change, it also poses challenges to societal unity and cohesion. Understanding the dynamics of identity politics is crucial for navigating the contemporary political landscape, recognizing its capacity to both unite and divide.

International Perspectives

The exploration of identity politics reveals a mosaic of manifestations across the globe, each uniquely influenced by the complex web of local contexts, histories, and societal structures. This diversity underscores the universal relevance of identity politics, while also highlighting the specific nuances that characterize its impact in different regions.

In India, for example, the intricate lattice of caste and religious identities forms a crucial backdrop to political discourse and action. The caste system, deeply ingrained in the socio-cultural fabric of the nation, intersects with religious affiliations to shape electoral strategies, social policies, and daily interactions. Political parties often tailor their campaigns to appeal to specific caste or religious groups, leveraging these identities to mobilize support and consolidate power. This dynamic, while fostering a sense of community and representation among certain groups, can also deepen historical fissures and perpetuate exclusion and inequality.

Brazil presents another compelling case study, where racial and socioeconomic identities intersect in ways that significantly influence political allegiances and activism. The legacy of colonialism and slavery has left indelible marks on Brazilian society, creating a complex hierarchy of racial and economic disparities. These disparities are mirrored in the political arena, where movements for racial equality and social justice vie for attention and reform against a backdrop of entrenched privilege and resistance. The mobilization around these identity-based issues has the potential to catalyze significant social change, yet it also faces challenges in bridging the vast divides that separate different segments of the population.

These international perspectives on identity politics illustrate the varied ways in which it can serve as both a tool for empowerment and a source of division. They underscore the importance of understanding the specific historical, cultural, and social contexts that shape identity politics in different regions, offering valuable insights into the challenges and opportunities inherent in addressing its complex dynamics. Through a comparative analysis, it becomes evident that while the manifestations of identity politics may differ globally, the underlying themes of recognition, representation, and equity are universally resonant.

Challenging the Narrative

To transcend the entrenched politics of division that pervade our global society, a comprehensive strategy focused on challenging divisive narratives and cultivating a culture rooted in empathy and understanding is essential. This approach necessitates empowering individuals with the skills and mindset to navigate the complex landscape of modern identity politics and social divisions. Central to this endeavor are critical thinking and media literacy, which equip individuals to critically analyze and question the oversimplified "us vs. them" narratives that frequently dominate media and political discourse. By fostering these skills, people can better recognize the nuanced and varied nature of social identities, seeing beyond binary categorizations to the rich tapestry of human experience that unites us all.

Encouraging dialogue and engagement across societal divides is another cornerstone of this strategy. Community initiatives, educational programs, and even personal conversations hold immense potential to bridge the gaps that separate us. By creating spaces where individuals from diverse backgrounds and perspectives can come together to share experiences, listen, and learn from one another, we can begin to dismantle the barriers of misunderstanding and prejudice that fuel division. These interactions, grounded in respect and genuine curiosity, can reveal the common values and shared aspirations that often go unnoticed in the heat of polarized debates.

Moreover, promoting a sense of common humanity is crucial in overcoming the divisive effects of identity politics. Recognizing that, despite our differences, we share fundamental needs, hopes, and fears, can help to soften the rigid boundaries that identity politics often constructs. This realization fosters empathy, allowing individuals to empathize with the struggles and perspectives of those who may seem vastly different on the surface.

Challenging the divisive narratives that underpin much of contemporary social and political life is no small task. It requires persistence, courage, and an unwavering commitment to the principles of empathy, understanding, and mutual respect. However, through concerted effort and collective action, it is possible to envision and work towards a future where division is replaced with dialogue, and where the politics of "us vs. them" gives way to a more inclusive and compassionate "we."

Promoting Inclusivity and Understanding

Promoting inclusivity and understanding in the face of division is not merely an idealistic pursuit but a practical endeavor with numerous successful precedents. Central to these successes are initiatives and approaches that emphasize shared experiences and common goals, effectively transcending the confines of narrow identity boundaries. By focusing on what unites rather than what divides, these efforts have shown significant promise in fostering a more cohesive and harmonious society.

Policies and practices aimed at inclusivity and diversity represent crucial tools in this ongoing effort. Affirmative action, for example, seeks to address historical inequalities and disparities by ensuring that marginalized groups have access to opportunities in education, employment, and other areas. While affirmative action has its critics, its role in promoting diversity and leveling the playing field cannot be underestimated. Similarly, multicultural education plays a pivotal role in broadening perspectives and fostering an appreciation for the rich elaborate network of human cultures and experiences. By exposing individuals to diverse viewpoints and histories, multicultural education challenges stereotypes and prejudices, laying the groundwork for a more inclusive society.

Another powerful mechanism for bridging divides is the celebration and dissemination of stories that highlight cooperation and mutual respect among diverse groups. These narratives serve as potent antidotes to divisive rhetoric, showcasing the tangible benefits and joys of unity in diversity. Whether it's communities coming together in times of crisis, joint initiatives between different cultural or religious groups, or individual stories of friendship and solidarity across divides, such examples provide compelling evidence of the potential for harmonious coexistence and mutual understanding.

In promoting inclusivity and understanding, it's essential to recognize that diversity is not just a challenge to be managed but a strength to be embraced. The variety of human experiences and perspectives is a wellspring of innovation, creativity, and resilience. By valuing and actively engaging with this diversity, societies can forge a path toward greater unity, where differences are celebrated and common humanity is the foundation for mutual respect and cooperation. Through concerted efforts to promote inclusive policies, educate for diversity, and highlight stories of collaboration, we can envision and work toward a world characterized by understanding and inclusivity, transcending the divisions that have long hindered our collective potential.

Lessons from History

      Lessons from history illuminate both the perilous consequences of unchecked division and the profound potential for reconciliation and unity. Historical episodes of sectarian violence, ethnic conflict, and social upheaval serve as stark reminders of what can occur when divisions are allowed to fester and escalate. Conversely, periods of reconciliation and collective movement towards equality and understanding highlight humanity's capacity for overcoming deep-seated divides.

The Civil Rights Movement in the United States offers a powerful case study in the dynamics of societal division and the pathways to overcoming it. This period was marked by intense racial segregation and discrimination, with African Americans and other minorities fighting for basic civil rights and equality. The movement, through its emphasis on nonviolent protest and civil disobedience, brought national and international attention to the injustices of racial segregation and discrimination. It demonstrated the impact of unified, peaceful action against systemic injustice, showing how deeply entrenched divisions can be challenged and ultimately transformed through persistence, courage, and solidarity.

Similarly, the reconciliation process in post-apartheid South Africa provides critical insights into healing the wounds of a deeply divided society. The establishment of the Truth and Reconciliation Commission (TRC) as a restorative justice body aimed to uncover the truth about past abuses and foster reconciliation between perpetrators and victims. This process highlighted the importance of acknowledging past injustices, offering forgiveness, and working collectively towards a shared future. It underscored the role of empathy, open dialogue, and mutual understanding in bridging historical divides and building a more inclusive society.

These historical examples underscore the challenges involved in overcoming divisions but also the transformative potential of concerted efforts toward reconciliation and unity. By examining these and other instances, we can glean valuable lessons on the importance of confronting divisive narratives, fostering inclusivity, and promoting understanding. These lessons, when applied to the contemporary context, can guide efforts to mitigate division and work towards a more unified and equitable society.

Future Implications

The current epoch, characterized by its intense negotiations between the forces of division and unity, stands at a pivotal crossroad in the annals of human history. Future historians might look back on these times as a defining moment when the fabric of global society was tested by the strains of identity politics and the "us vs. them" mentality. The narratives that emerge from our era will likely underscore the critical importance of addressing the foundational causes of division—such as systemic inequality, entrenched injustice, and pervasive fear—while also highlighting the indispensable roles played by empathy, dialogue, and inclusivity in weaving a more cohesive social tapestry.

As we navigate through the complexities of identity and division, our actions and choices bear the weight of future implications. The potential for a more unified world hinges on our ability to learn from history, to draw upon the lessons of movements that have bridged vast divides and to apply those lessons to the present context. The Civil Rights Movement in the United States, the reconciliation process in post-apartheid South Africa, and other similar historical episodes offer invaluable insights into the dual nature of humanity's capacity for conflict and reconciliation. These examples demonstrate that, even in the face of deep-seated divisions, paths toward understanding and unity can be forged.

Embracing a culture of empathy allows us to see beyond the immediate confines of our own experiences and identities, fostering a deeper appreciation for the diversity of human life. Critical thinking and media literacy equip us with the tools to dissect and challenge divisive narratives, promoting a more nuanced and inclusive understanding of the issues at hand. Moreover, policies and practices that celebrate diversity and encourage inclusivity, such as affirmative action and multicultural education, contribute to the dismantling of barriers that segregate us.

The journey forward, away from the entrenched politics of "us vs. them," toward a future where diversity is not a source of division but a wellspring of strength and unity, is fraught with challenges. Yet, it is also ripe with opportunity. By leveraging the lessons of history, embracing the values of empathy and inclusivity, and fostering dialogue across our differences, we can chart a course toward a more harmonious and unified global community. The choices we make today will not only shape the narratives of tomorrow but also offer hope for a world where unity in diversity prevails as our greatest collective strength.

Part III: Governance and Political Shortcomings

Chapter 9: Democracy's Decline - The Rise of Authoritarianism

At the dawn of the 21st century, the future of democracy seemed promising, buoyed by the fall of authoritarian regimes and the global spread of democratic ideals. However, the ensuing decades have witnessed a disconcerting shift, as signs of democracy's decline emerge across the world, from established powers to burgeoning democracies. This chapter aims to unravel the complex phenomenon of democratic backsliding and the concurrent rise of authoritarian governance, exploring its causes, manifestations, and the potential pathways for reversing this trend. By understanding these dynamics, we confront not just a political challenge, but a pivotal question of our time: How can the ideals of democracy be preserved and revitalized in the face of growing authoritarian tendencies?

Historical Context

The aftermath of the Second World War marked the beginning of an unparalleled era of democratic expansion. Nations shattered by war sought to rebuild themselves on the principles of freedom, justice, and democracy. The international community, eager to prevent the repetition of such devastating conflict, established institutions and norms aimed at promoting peace, cooperation, and democratic governance. This era witnessed the formation of the United Nations, the adoption of the Universal Declaration of Human Rights, and the implementation of the Marshall Plan, all of which contributed significantly to the global spread of democratic ideals.

The end of the Cold War served as a catalyst for democracy's rapid spread, particularly with the democratization of former Soviet states. The dissolution of the USSR in 1991 not only signified the end of bipolar world order but also removed a significant barrier to the adoption of democratic governance in many countries. This period saw the emergence of new democracies in Eastern Europe, the Baltic states, and elsewhere, marking what some considered the zenith of democracy's global influence. The widespread belief in the triumph of liberal democracy was epitomized by Francis Fukuyama's "end of history" thesis, which argued that liberal democracy might represent the endpoint of mankind's ideological evolution and the final form of human government.

The initial optimism of the late 20th century has been increasingly challenged by the realities of the 21st century. Despite the significant gains made, democratic institutions across the globe have shown signs of vulnerability. Established democracies have grappled with internal divisions, political polarization, and a growing distrust in public institutions. At the same time, external threats, such as cyber warfare and disinformation campaigns, have targeted the integrity of democratic processes.

The rise of authoritarian leaders in countries with young or fragile democracies has demonstrated the ease with which democratic backsliding can occur. These leaders, often exploiting economic grievances or societal divisions, have gradually eroded the checks and balances essential to a healthy democracy. This has led to an alarming trend where the line between democratic and authoritarian governance blurs, challenging the notion that the spread of democracy is irreversible or that it represents the ultimate form of governance.

While the post-World War II era laid the foundation for a dramatic expansion of democratic governance, the subsequent decades have revealed the complexities and challenges of maintaining and expanding upon those gains. The optimism of the "end of history" has given way to a more nuanced understanding that democracy is not a given but a system that requires constant nurturing, defense, and revitalization.

Current Landscape

Democratic backsliding emerges as a phenomenon both subtle and insidious, marking a departure from the overt seizures of power seen in military coups or revolutions. This form of decline is characterized by the erosion of the very norms and practices that underpin democratic systems, all while operating within the guise of democracy itself. What makes backsliding particularly pernicious is its incremental nature; elected leaders, under the mantle of legitimacy provided by their office, gradually undermine the institutions meant to check their power. This slow-motion erosion often escapes immediate detection, allowing it to advance largely unchallenged until the damage becomes unmistakably clear.

One of the most troubling aspects of democratic backsliding is its ability to blur the distinctions between democratic and authoritarian governance. As elected leaders adopt increasingly authoritarian practices—such as eroding the independence of the judiciary, restricting the press, manipulating electoral processes, and sidelining political opposition—they do so without overtly abandoning the trappings of democracy. Elections may still occur, but their fairness is compromised; media remains, but under heavy censorship or influence; opposition exists, but is systematically weakened or discredited. This hybridization of practices not only challenges the resilience of democratic systems but also confounds the international community's ability to respond effectively.

The current landscape of democratic governance is thus one of contradiction and challenge. On one hand, the formal structures of democracy often remain intact, providing a veneer of legitimacy to those who seek to undermine it from within. On the other, the substance of democracy—free elections, rule of law, protection of rights and freedoms—is increasingly under threat. This presents a formidable challenge to the resilience of democratic systems, requiring not just vigilance but a reinvigoration of democratic norms and institutions. The task is no less than to identify and counteract the tactics of backsliding, ensuring that democracy's form and function are preserved against the encroachment of authoritarian practices.

Economic Factors

Economic instability has emerged as a critical catalyst for the erosion of faith in democratic institutions. The seismic shocks of the 2008 financial crisis and its resultant global recession serve as a stark example of this dynamic. These economic upheavals exposed the vulnerabilities of individuals and communities, leading to a profound crisis of confidence in the ability of democratic systems to safeguard economic security and prosperity. The aftermath of these events saw a marked increase in the susceptibility of populations to the allure of populist and authoritarian narratives, which often promise swift and decisive action to rectify economic grievances.

Compounding the issue of economic instability is the specter of growing inequality. The benefits of globalization, while substantial, have been unevenly distributed, often enriching a cosmopolitan elite while leaving wide swathes of the global population feeling marginalized and disenfranchised. This perception of inequality has become a fertile ground for populist backlash, challenging the prevailing consensus on the virtues of liberal democracy and free market principles. Populist leaders capitalize on this discontent, framing globalization and its champions within the democratic establishment as the root causes of the populace's economic woes.

The intertwining of economic instability and inequality with democratic backsliding cannot be overstated. Economic grievances provide a powerful tool for leaders with authoritarian tendencies to undermine democratic norms and institutions. By promising economic revival, often through illiberal means, these leaders can garner significant support, enabling them to enact policies that concentrate power and suppress dissent. This dynamic underscores a vicious cycle wherein economic dissatisfaction fuels authoritarian practices, which in turn further undermines the democratic fabric.

The challenge for democracies in the face of these economic pressures is to find ways to address legitimate economic grievances without sacrificing democratic principles. This entails pursuing policies that promote economic fairness and security, such as progressive taxation, social safety nets, and investment in education and infrastructure. By demonstrating that democratic systems can deliver tangible economic benefits, democracies can begin to rebuild trust and counteract the appeal of populist and authoritarian alternatives.

Economic factors play a pivotal role in shaping the landscape of global democracy. Economic instability and growing inequality not only fuel disillusionment with democratic systems but also enhance the appeal of populist and authoritarian narratives. Addressing these economic challenges is crucial for the preservation and revitalization of democratic governance.

Political and Social Factors

The fabric of social cohesion, once the bedrock of democratic societies, has become increasingly frayed in the contemporary era. This erosion is attributable, in part, to rapid cultural and demographic changes that have outpaced societies' ability to adapt and integrate new realities. As a consequence, the sense of a unified community, with shared values and objectives, has diminished, giving way to heightened social divisions. These divisions often manifest along lines of ethnicity, religion, socioeconomic status, and political affiliation, creating fissures that populist leaders are adept at exploiting. By emphasizing these divisions, they can mobilize support through divisive rhetoric, framing their ascent to power as a victory for the marginalized or misrepresented.

Closely tied to the erosion of social cohesion is the ascendancy of identity politics. This phenomenon has seen political allegiance and activism increasingly organized around particular identity markers, rather than broad-based ideological or policy platforms. While identity politics can play a vital role in highlighting and addressing the grievances of marginalized groups, it can also contribute to deepening social divisions when leveraged by opportunistic leaders. Such leaders may amplify identity-based grievances to create an 'us versus them' mentality, undermining the inclusive dialogue and compromise essential for democratic governance.

The digital age, for all its benefits, has introduced a formidable challenge to democratic systems through the proliferation of misinformation. Social media platforms, in particular, have facilitated the rapid spread of false information, creating a post-truth environment where objective facts are often obscured or disregarded. This phenomenon undermines the foundation of an informed electorate, which is essential for the functioning of democracy. Voters deprived of reliable information are more susceptible to manipulation, making it easier for unscrupulous leaders to exploit misinformation for political gain.

The undermining of an informed electorate through misinformation not only challenges the efficacy of democratic decision-making but also erodes trust in democratic institutions and processes. When facts become contested territory, and the truth is viewed through the lens of political allegiance, the capacity for constructive public discourse and consensus-building is significantly diminished. This environment fosters cynicism and disengagement among citizens, further weakening the foundations of democratic governance.

Addressing Political and Social Challenges

The interplay between the political and social factors outlined above presents a complex challenge to the sustainability of democratic systems. Addressing these challenges requires a versatile approach, including efforts to rebuild social cohesion, promote inclusive dialogue, and combat misinformation. Democracies must invest in education and critical thinking skills, encourage cross-cultural and interfaith understanding, and enhance the transparency and accountability of information sources. By tackling these issues, democracies can work to repair the social fabric and reaffirm the principles of informed, inclusive governance that underpin democratic society.

The political and social landscape for democracy is marked by significant challenges, including the erosion of social cohesion, the rise of identity politics, and the spread of misinformation. Addressing these challenges is essential for preserving the integrity and functionality of democratic systems in the modern era.

Erosion of Democratic Institutions

The foundation of a robust democracy lies in its institutions and the checks and balances that prevent the concentration of power. However, this foundation is precisely what is targeted in the process of democratic erosion. Authoritarian leaders, often emerging from within the democratic system, exploit its vulnerabilities to gradually undermine these vital safeguards. Their approach is methodical, leveraging the very mechanisms of democracy to weaken it incrementally, making it difficult to mount effective resistance until the damage becomes severe.

One of the primary targets in the erosion of democratic institutions is the independence of the judiciary. Authoritarian leaders seek to control the courts to ensure legal challenges against their actions are nullified. Packing courts with loyalists is a common strategy, transforming the judiciary from a check on executive power into an enabler of authoritarian ambitions. This not only undermines the rule of law but also erodes public trust in the justice system as a fair arbitrator.

Another critical institution under attack in the erosion of democracies is the free press. Authoritarian leaders recognize the power of the media to shape public opinion and uncover abuses of power. In response, they impose restrictions on media freedom, through direct censorship, regulatory constraints, or economic pressures to silence critical voices. The free press, essential for an informed electorate and a cornerstone of democracy, is thus compromised, limiting its ability to hold power to account.

The manipulation of electoral processes is a direct assault on the heart of democracy. By altering electoral laws, redrawing voting districts to their advantage, or implementing measures that suppress voter turnout among opposition supporters, authoritarian leaders can entrench their power. These tactics not only skew the democratic process in their favor but also erode the fundamental principle of democracy: that power derives from the consent of the governed, freely expressed through fair and competitive elections.

The cumulative effect of these strategies is the gradual, often imperceptible, transformation of democratic systems into authoritarian regimes. What makes this process particularly insidious is its incremental nature, which allows leaders to claim legitimacy from democratic origins even as they dismantle democracy's core features. The erosion of democratic institutions represents not just a power grab by authoritarian leaders but a profound threat to the principles of freedom, justice, and equality that define democratic governance.

The erosion of democratic institutions is a deliberate process, exploited by authoritarian leaders to undermine the checks and balances that are the hallmark of democratic systems. By attacking the judiciary, limiting press freedom, and manipulating electoral processes, they erode the very foundations of democracy from within, posing a significant challenge to democratic resilience and the protection of fundamental freedoms.

Suppression of Dissent and Civil Liberties

The ascension of authoritarian regimes often heralds a period of heightened suppression of dissent and a systematic erosion of civil liberties, central to the consolidation of their control. This repression manifests in various forms, from the subtler use of legal instruments to silence critics and opposition figures, to more blatant acts of intimidation and violence against those who dare to challenge the status quo. The strategic manipulation of information and public discourse plays a crucial role in reinforcing authoritarian dominance. Through mechanisms of censorship and the propagation of state-sanctioned propaganda, these regimes seek to monopolize the narrative, ensuring that only their version of truth prevails.

One of the most insidious tactics employed by authoritarian leaders is the use of legal mechanisms to suppress dissent. Laws ostensibly aimed at protecting national security, combating terrorism, or preserving public order are frequently repurposed to target political opponents, activists, and journalists. The effect is twofold: it eliminates key voices of opposition and instills a pervasive fear of reprisal among the populace, significantly dampening public discourse and opposition activities.

Beyond legal frameworks, authoritarian regimes often resort to direct intimidation and violence against those perceived as threats. This can range from arbitrary arrests and detention to physical assaults and even extrajudicial killings. Such actions send a clear message about the costs of dissent, aiming to break the spirit of resistance and solidarity among critics and the broader public.

A hallmark of authoritarian control is the domination over information and public discourse. Through censorship, the suppression of independent media, and the dissemination of propaganda, authoritarian regimes strive to eliminate dissenting views and promote an alternative reality that supports their governance. This control extends into the digital realm, where internet censorship and surveillance are used to monitor dissent and manipulate online narratives.

The suppression of dissent is intrinsically linked to the erosion of fundamental freedoms, including the rights to assembly, speech, and political participation. By restricting these rights, authoritarian regimes aim to prevent the formation of opposition movements, stifle public protests, and ensure that electoral processes remain firmly under their control. The curtailing of these freedoms not only undermines the bedrock of democratic engagement but also signals a move towards a more closed and oppressive society.

The rise of authoritarianism brings with it an increased suppression of dissent and a deliberate erosion of civil liberties. Through a combination of legal mechanisms, intimidation, control over information, and the curtailing of fundamental freedoms, authoritarian leaders seek to solidify their grip on power, often at the expense of the democratic fabric of society. Addressing these challenges requires a concerted effort from democratic institutions, civil society, and the international community to safeguard the principles of freedom and democracy.

Western Democracies

The political landscape within Western democracies has undergone significant shifts in recent years, revealing underlying vulnerabilities and stresses in their democratic systems. The rise of populist movements across these nations has brought to the fore challenges that threaten the fabric of democracy, utilizing narratives that often appeal to national sovereignty, economic disenfranchisement, and cultural identity.

A prime example of the impact of populist movements in Western democracies is the United Kingdom's Brexit referendum. The decision to leave the European Union was significantly influenced by populist rhetoric that emphasized national sovereignty and control over immigration. This referendum not only highlighted the potent appeal of populist arguments but also exposed deep societal divisions, raising questions about the role of misinformation in democratic processes and the potential consequences of majoritarian decision-making on complex issues.

In the United States, increasing political polarization has become a pervasive concern, with the gap between the two main political parties widening on numerous issues. This polarization is not just ideological but also reflects deeper divisions within American society, relating to race, economic status, and cultural values. Challenges to democratic norms have become more apparent, with instances of questioning the legitimacy of electoral processes, undermining the free press, and eroding the independence of the judiciary. These developments have sparked debates about the resilience of American democratic institutions and the need for reforms to ensure their integrity and functionality.

In both the United Kingdom and the United States, as well as other Western democracies, populist movements have successfully tapped into a sense of disenchantment among certain segments of the population. By framing their narratives around issues such as national identity, economic insecurity, and disillusionment with the political elite, these movements have been able to challenge traditional party structures and question the efficacy of democratic governance models. This has led to a reevaluation of how democracies can better address the needs and concerns of all citizens, ensuring that the benefits of democratic governance are widely shared and that the institutions themselves remain robust and responsive.

The experiences of Western democracies with populist movements underscore the need for continual vigilance and adaptation within democratic systems. Addressing the root causes of societal divisions, improving the inclusivity and transparency of democratic processes, and reinforcing the norms and institutions that underpin democracy are crucial steps in safeguarding democratic resilience. As Western democracies navigate these challenges, the lessons learned will be invaluable for democratic societies worldwide, highlighting the importance of unity, dialogue, and a steadfast commitment to democratic principles.

Western democracies are at a crossroads, confronted with the dual challenge of responding to the rise of populist movements and ensuring the continued health and vitality of their democratic systems. The way forward requires a nuanced understanding of the factors driving societal divisions and a concerted effort to reinforce the democratic compact between governments and their citizens.

Emerging Democracies

The landscape of emerging democracies provides a stark illustration of how the process of democratic backsliding can unfold, even in countries with histories of democratic elections and governance. Nations such as Türkiye and Hungary serve as critical case studies in understanding the methods and consequences of this phenomenon, where democratically elected leaders systematically dismantle democratic structures to consolidate their power.

In Türkiye, President Recep Tayyip Erdogan's tenure has been marked by a significant shift towards authoritarian governance, despite the country's previous strides towards democratic consolidation. Erdogan has skillfully utilized a combination of legal reforms, constitutional changes, and crackdowns on dissent to expand his presidential authority and weaken the opposition. Following a failed coup attempt in 2016, the government launched widespread purges against perceived enemies, targeting not only military personnel but also academics, journalists, and civil society members, severely restricting freedom of speech and assembly. These actions, coupled with the consolidation of control over the media and alterations to the electoral system, have significantly eroded democratic checks and balances in Türkiye, raising serious concerns about the future of its democratic institutions.

Hungary presents another alarming example of democratic backsliding, with Prime Minister Viktor Orban at the helm of the country's drift towards authoritarianism. Since coming to power, Orban and his Fidesz party have systematically dismantled Hungary's democratic frameworks, asserting control over the judiciary, the media, and other key institutions to entrench their power. The government has introduced a series of constitutional and legal changes that have effectively neutered the opposition, curtailed civil liberties, and manipulated electoral laws to favor the ruling party. Orban's government has also been accused of eroding media freedom by bringing news outlets under the control of government allies, thereby stifling critical voices and promoting a pro-government narrative.

Both Türkiye and Hungary highlight a pattern of democratic erosion that has become increasingly prevalent in emerging democracies around the world. These leaders leverage their initial democratic legitimacy to systematically dismantle the institutions designed to check their power. The erosion of democratic norms and institutions in these countries is not the result of sudden coups or revolutions but rather a gradual process facilitated by the manipulation of legal and constitutional frameworks.

The international community's response to the erosion of democracy in Türkiye, Hungary, and similar contexts has been mixed, with concerns often expressed through diplomatic channels, economic sanctions, or public criticism. However, the effectiveness of these measures in reversing democratic backsliding remains uncertain. The path forward for emerging democracies facing similar challenges lies in the resilience of civil society, the international reinforcement of democratic norms, and the cultivation of a political culture that prioritizes democratic governance and human rights.

The experiences of Türkiye and Hungary with democratic backsliding provide valuable lessons on the fragility of democratic institutions and the need for constant vigilance to protect them. The consolidation of power by democratically elected leaders through the erosion of democratic checks and balances is a trend that poses a significant challenge to the global democratic order, necessitating a concerted and unified response to uphold the principles of democracy and freedom.

Strengthening Institutions

In the face of rising authoritarian tendencies and democratic backsliding, fortifying democratic institutions emerges as a paramount strategy. The essence of democracy lies in its institutions, which are designed to uphold the rule of law, ensure accountability, and protect individual liberties. Strengthening these institutions is essential for countering the erosion of democratic norms and preventing the consolidation of power by authoritarian leaders. This involves a strategic approach, focusing on the judiciary, the media, and the electoral process, among other areas.

The judiciary plays a crucial role in maintaining democracy by ensuring that all actions by the government and its officials are in accordance with the law. Strengthening the judiciary involves ensuring its independence from the executive and legislative branches of government, allowing it to serve as an effective check on power. This can be achieved through measures such as secure tenure for judges, transparent and merit-based appointment processes, and adequate resourcing and training.

A free and independent press is vital for a healthy democracy, serving as a watchdog that holds power to account and provides citizens with the information necessary to make informed decisions. Protecting press freedom requires legal safeguards against censorship and intimidation, as well as support for journalistic integrity and independence. This includes the decriminalization of defamation, protection of sources, and the promotion of media diversity to prevent monopolies.

The integrity of electoral processes is fundamental to the legitimacy of democracy. Ensuring fair and free elections involves protecting against manipulation, interference, and fraud. Measures to safeguard electoral integrity include the establishment of independent electoral commissions, the implementation of transparent voting and counting procedures, and the provision of adequate security and monitoring. Additionally, addressing issues such as gerrymandering and ensuring equitable media access for all political parties are vital steps toward fair electoral competition.

International cooperation and support from democratic nations are invaluable in bolstering democratic institutions in countries under authoritarian pressures. This can take the form of diplomatic engagement, technical assistance, and the sharing of best practices. International organizations and democracies can offer support for judicial reform, media freedom, and electoral integrity through various means, including monitoring missions, capacity building, and financial aid.

The strengthening of democratic institutions is a critical defense against the advance of authoritarianism. By ensuring the independence of the judiciary, protecting freedom of the press, and safeguarding electoral integrity, democracies can build resilience against internal and external threats. International cooperation plays a crucial role in supporting these efforts, highlighting the interconnectedness of democratic societies in the global struggle for freedom and governance by the people. In this endeavor, the collective action of democratic nations and international institutions can provide the necessary support and solidarity to those fighting to maintain and strengthen their democratic systems.

Civic Engagement and Education

In the contemporary struggle against the decline of democracy and the ascent of authoritarianism, the empowerment of the citizenry through civic engagement and education stands out as a cornerstone strategy. The cultivation of a public that not only values but actively defends democratic principles is fundamental to the health and resilience of democratic systems. This section looks into the pivotal roles of civic participation and educational initiatives in fostering a robust democratic society, underscoring their importance in the broader effort to combat democratic backsliding.

Civic engagement encompasses a wide range of activities through which individuals participate in the life of their community and the governance of their country. This includes not just voting, which is the most basic act of democratic participation, but also engagement in civil society organizations, participation in protests and public demonstrations, and involvement in local governance and community projects. Encouraging such active participation is crucial for sustaining a vibrant civil society capable of holding authorities to account. It ensures that the voices of citizens are heard and that those in power remain responsive to the needs and concerns of the public they serve.

Parallel to the promotion of civic engagement is the critical role of education in preparing citizens to effectively participate in and navigate the complexities of the political landscape. An education that emphasizes critical thinking and media literacy equips individuals with the tools necessary to critically evaluate information, discern truth from misinformation, and make informed decisions. Furthermore, incorporating democratic values into the educational curriculum instills a deep appreciation for the principles of freedom, equality, and justice that are the bedrock of democratic societies. Such education fosters a sense of civic responsibility and a commitment to the common good, which are essential for the defense of democracy.

The unsettling trend of democracy's decline and the concurrent rise of authoritarianism underscore the urgency of fostering widespread civic engagement and robust educational frameworks. These initiatives serve not only as bulwarks against the erosion of democratic norms but also as engines for the renewal and strengthening of democratic societies. By cultivating a well-informed and actively engaged citizenry, societies can better confront the authoritarian challenge, ensuring that democratic institutions remain strong and that the ideals of democracy are preserved for future generations.

As this chapter has outlined, the battle against the tide of authoritarianism and for the soul of democracy is fought on many fronts. The analysis of the causes, manifestations, and case studies of democratic decline provides a clear picture of the challenges faced. Yet, it also offers hope, presenting actionable pathways toward resilience. Strengthening democratic institutions, fostering civic engagement, and prioritizing education in democratic values are strategies that, together, form a powerful blueprint for resistance and renewal. In this critical moment, the collective efforts to promote engagement and education are not just strategies for preservation but acts of defiance against authoritarian encroachment, safeguarding the democratic ideals that underpin a free and just society.

Chapter 10 Post-Truth Politics - The Assault on Fact and Reason

In recent years, political discourse across the globe has witnessed a paradigmatic shift toward what is now commonly referred to as post-truth politics. This phenomenon, characterized by the relative sidelining of facts and the elevation of emotional appeal and personal belief, poses profound challenges to the foundational principles of democratic discourse and governance. The term "post-truth" itself, Oxford Dictionaries' 2016 Word of the Year, encapsulates the growing disregard for factual accuracy in shaping public opinion and political decisions. This chapter looks into the origins, mechanisms, and societal impacts of post-truth politics, seeking to understand how we arrived at this juncture and what can be done to navigate a path forward, where truth and reason reclaim their central place in political life.

Historical Context

The phenomenon of post-truth politics, while appearing as a distinctly 21st-century dilemma, has its underpinnings deeply rooted in the latter half of the 20th century. This period marked the beginning of a gradual shift towards a more skeptical view of objective truths, fueled in part by philosophical movements that questioned the very existence of universal truths. These philosophical debates laid the groundwork for a cultural and political landscape more receptive to relative truths, where facts could be seen as pliable and subject to interpretation based on one's perspective or political allegiance.

Parallel to this philosophical skepticism was the transformation of the media and communication landscape. The advent of the internet and the subsequent rise of social media platforms revolutionized the way information was created, shared, and consumed. This digital revolution brought with it unprecedented access to information but also presented new challenges in discerning the quality and veracity of that information. The sheer volume of content available online made it increasingly difficult for individuals to navigate the information landscape, leading to the proliferation of misinformation and the development of ideological echo chambers.

Social media, in particular, has played a pivotal role in this transition. Algorithms designed to engage users by feeding them content that aligns with their existing beliefs have further entrenched ideological divisions, creating silos of information that rarely intersect. This digital echo chamber effect has been instrumental in the rise of post-truth politics, where narratives often matter more than factual accuracy, and emotional resonance can trump objective analysis.

The late 20th century saw significant political and cultural shifts that contributed to the fertile ground for post-truth politics. The end of the Cold War and the advent of globalization brought about a reevaluation of national identities and truths, as societies grappled with the complexities of a more interconnected world. These changes, combined with growing skepticism toward traditional institutions and media, fostered an environment where subjective truths and personal beliefs could easily challenge established facts.

The culmination of these factors—philosophical skepticism, the transformation of media, the rise of social media, and significant political and cultural shifts—has contributed to the emergence and entrenchment of post-truth politics. This historical context is crucial in understanding how we arrived at the current juncture, where objective facts often take a backseat to personal beliefs and emotional appeal in political discourse. As we move forward, acknowledging and addressing the roots of this phenomenon will be key in navigating the challenges it presents to democratic discourse and governance.

Characteristics and Causes

Post-truth politics is distinguished by several key features that collectively mark its departure from traditional political discourse. These include:

- Blurring of Fact and Opinion: In the realm of post-truth politics, the distinction between objective facts and subjective opinions becomes increasingly indistinct. Statements and claims, regardless of their factual accuracy, are often presented with the same level of authority as well-established truths.

- Emotional and Belief-Based Appeals: Emotional resonance and personal beliefs take precedence over empirical evidence and logical reasoning. Political arguments and campaigns are structured to appeal directly to the emotions of the audience, often invoking fear, anger, or nostalgia, rather than presenting objective evidence or logical argumentation.

- Contingent Truths: The concept of truth becomes fluid and adaptable, contingent upon social or political affiliations. Facts are no longer universally acknowledged truths but are seen through the lens of group identity, leading to a scenario where the same fact can be interpreted in diametrically opposed ways by different groups.

The ascent of post-truth politics is not attributable to a single cause but rather to a confluence of factors that have disrupted traditional modes of political discourse and information sharing:

- Digital Media Evolution: The advent and proliferation of digital media have fundamentally transformed how information is disseminated and consumed. The fragmentation of the media landscape means that individuals have access to an overwhelming array of sources, many of which tailor content to specific interests or viewpoints, reinforcing preexisting beliefs and biases.

- Political Polarization: Increasing political polarization has created an environment where political figures and parties are incentivized to appeal to the emotional and irrational instincts of their base rather than engaging in rational discourse. This polarization encourages a binary view of political issues, where the middle ground is eroded, and compromise becomes increasingly difficult.

- Erosion of Trust in Traditional Media and Institutions: A significant decline in public trust in traditional media outlets and expert institutions has created a vacuum that is readily exploited by purveyors of post-truth narratives. As trust in these traditional arbiters of truth wanes, individuals turn to alternative sources of information, often those that confirm their preexisting views, regardless of factual accuracy.

- The Role of Social Media: Social media platforms have exacerbated the spread of misinformation and the formation of echo chambers. Their algorithms prioritize content that engages users, regardless of its veracity, leading to the rapid dissemination of falsehoods and the reinforcement of ideological divides.

Together, these characteristics and causes paint a picture of a political landscape in which the objective truth is increasingly sidelined in favor of narratives that appeal to emotion, belief, and identity. This shift poses significant challenges to democratic discourse, necessitating a critical examination of how societies can address the underlying causes and mitigate the effects of post-truth politics.

The Role of Social Media

Social media platforms have emerged as central actors in the ecosystem of post-truth politics. Their unique structure and functioning have significantly contributed to the shaping of a political and social environment where factual accuracy often takes a backseat to emotional engagement and ideological reaffirmation. The influence of social media in this context can be understood through two primary mechanisms: content prioritization algorithms and the facilitation of echo chambers.

The algorithms that underpin content distribution on social media platforms are designed to maximize user engagement. These algorithms use various metrics, such as likes, shares, and the amount of time spent on content, to determine what users find engaging. Unfortunately, sensationalist and emotionally charged misinformation often generates high levels of engagement, leading these algorithms to prioritize such content in users' feeds. This dynamic creates a feedback loop where the most engaging (rather than the most accurate) information is more widely disseminated, amplifying the reach and impact of misinformation.

Social media platforms also contribute to the formation of echo chambers, digital environments where users are primarily exposed to opinions and information that reinforce their existing beliefs. Through the use of personalized content feeds, users are more likely to encounter information that aligns with their views, while content that challenges or contradicts these views is filtered out. This mechanism, combined with user-driven content selection (such as following certain pages or individuals), means that social media users can end up in highly curated informational bubbles. These echo chambers exacerbate societal divisions by reducing the opportunities for exposure to diverse perspectives and making it more challenging for users to encounter or engage with conflicting information.

The role of social media in post-truth politics has profound implications for public discourse and democracy. By prioritizing emotionally engaging misinformation and facilitating the creation of ideologically homogeneous echo chambers, social media platforms contribute to a fragmented public square where shared realities are increasingly rare. This fragmentation poses significant challenges to democratic governance, which relies on a well-informed electorate and robust, fact-based public discourse.

In addressing the challenges posed by social media in the post-truth era, it is essential to consider reforms and innovations that can mitigate these dynamics. Potential measures include algorithmic transparency, the promotion of digital literacy, and the development of tools and policies designed to curb the spread of misinformation while encouraging exposure to a broader range of perspectives.

Cognitive Biases and Emotional Manipulation

At the heart of post-truth politics lies the manipulation of innate human psychological tendencies—cognitive biases—that significantly influence how information is processed and perceived. These biases, deeply embedded in human cognition, play a crucial role in shaping individuals' responses to political messaging, often to the detriment of factual accuracy and rational discourse. Understanding these biases is key to comprehending the effectiveness of post-truth strategies and the challenges they pose to informed public engagement.

One of the most pervasive cognitive biases is confirmation bias, which drives individuals to seek out, interpret, and remember information in a way that confirms their preexisting beliefs and opinions. This bias means that people are more likely to give credence to information, accurate or not, that aligns with their worldview and to discount evidence that contradicts it. In the realm of post-truth politics, confirmation bias is exploited to reinforce existing narratives and ideologies, making it easier for misleading or false information to gain traction if it resonates with an individual's preconceived notions.

Cognitive dissonance, another fundamental psychological phenomenon, occurs when individuals are presented with information that conflicts with their existing beliefs or opinions, leading to discomfort or distress. To alleviate this discomfort, people often reject, rationalize, or ignore the conflicting information, a tendency that post-truth politics manipulates by creating narratives that are emotionally comforting or affirming, even if they are factually inaccurate. Politicians and interest groups leverage this discomfort by offering simplistic, emotionally appealing answers to complex issues, sidestepping the need for factual accuracy.

The emotional manipulation inherent in post-truth politics capitalizes on these cognitive biases by crafting messages that resonate deeply on an emotional level, often invoking fear, anger, nostalgia, or hope, rather than appealing to logic or reason. These messages are designed to create a strong emotional response that overrides rational evaluation of the accuracy or veracity of the information presented. By tapping into emotions, post-truth practitioners can more effectively propagate narratives that support their political objectives, regardless of their relationship to the truth.

The strategic exploitation of cognitive biases and emotional responses is a powerful tool in the arsenal of post-truth politics. It allows politicians and interest groups to craft messages that bypass rational scrutiny and directly influence individuals' perceptions and beliefs. This exploitation of psychological vulnerabilities underscores the challenges faced by societies in fostering a political culture that prioritizes factual accuracy and rational discourse over emotional manipulation and ideological confirmation.

Addressing the impact of cognitive biases and emotional manipulation in post-truth politics requires integrated approach, including enhancing critical thinking and media literacy among the public, fostering environments that encourage exposure to diverse perspectives, and developing strategies to counteract the emotional appeal of misinformation. By understanding and mitigating the influence of these psychological dynamics, it is possible to strengthen the foundations of democratic discourse and governance in the face of post-truth challenges.

Undermining Trust and Consensus

The rise of post-truth politics marks a significant shift in the landscape of public discourse, with profound implications for societal trust and cohesion. By prioritizing emotional resonance over factual accuracy, post-truth politics not only challenges the veracity of specific claims or narratives but also fundamentally undermines the trust upon which democratic societies are built. This erosion of trust extends to various traditional institutions that have historically served as pillars of factual information and rational debate, including the media, the scientific community, and government entities.

One of the most significant impacts of post-truth politics is the diminishing public trust in established institutions. As the line between fact and fiction blurs, individuals become increasingly skeptical of the information presented by these institutions, viewing it through the lens of political bias or as part of a broader narrative agenda. This skepticism is exacerbated by deliberate campaigns of misinformation and disinformation aimed at discrediting these institutions, further eroding their perceived credibility and authority.

The decline in trust is particularly concerning in the context of the media and the scientific community, institutions that rely on public confidence to fulfill their roles effectively. In the case of the media, accusations of bias and "fake news" have led to a fragmentation of the information landscape, with individuals seeking out alternative sources that align with their views, regardless of factual accuracy. Similarly, the scientific community faces challenges in communicating complex, often inconvenient truths in an environment where empirical evidence is contested by ideologically driven counter-narratives.

The undermining of trust in traditional institutions contributes to a broader societal fragmentation, dividing communities along lines of competing realities. This division is not merely ideological but epistemological, with different segments of society operating on fundamentally different understandings of truth and fact. This fragmentation poses significant challenges to achieving consensus on critical issues, as agreement on solutions presupposes a shared recognition of the problems themselves.

The difficulty in reaching consensus is particularly evident in debates on public health, climate change, and other areas where scientific knowledge and policy intersect. In these domains, the rejection of established facts in favor of alternative narratives has complicated efforts to formulate and implement effective responses, often turning issues that are fundamentally scientific or technical in nature into arenas of political conflict.

Addressing the challenges posed by the undermining of trust and consensus in the age of post-truth politics requires concerted efforts across multiple fronts. Rebuilding trust in institutions necessitates a commitment to transparency, accountability, and the promotion of media literacy among the public. It also involves the creation of spaces for dialogue that transcend ideological divides, facilitating engagement with diverse perspectives and fostering a culture of critical thinking and skepticism towards misinformation.

Ultimately, the struggle against post-truth politics and its corrosive effects on society is not just about defending factual accuracy but about reaffirming the foundational principles of democratic discourse and governance. By working to restore trust and build consensus, societies can navigate the challenges of the post-truth era and move towards a more informed and cohesive future.

Challenges to Policy Making and Governance

In the era of post-truth politics, the landscape of governance and policy-making faces profound challenges. The shift away from evidence-based decision-making towards an environment where populist sentiment and misinformation exert significant influence poses substantial risks to the development and implementation of effective policy measures. This transition not only complicates the process of governance but also threatens the long-term well-being and stability of societies.

One of the key challenges in this new political landscape is the rise of populism, characterized by appeals to the emotions and prejudices of the public rather than reasoned debate and empirical evidence. Populist movements often leverage misinformation to gain support, crafting narratives that resonate with the fears and desires of the populace while disregarding the complexities of policy issues. This reliance on misinformation can lead to policy decisions that are not grounded in reality, potentially resulting in ineffective or counterproductive outcomes.

The spread of misinformation also undermines the role of expertise and scientific evidence in policy-making. As public discourse becomes increasingly dominated by unverified or false information, the voices of experts and scientists are drowned out or dismissed, reducing the capacity for informed decision-making. This rejection of expert opinion is particularly detrimental in areas such as public health, environmental policy, and economic strategy, where specialized knowledge is crucial for developing effective and sustainable solutions.

The post-truth political climate further stifles constructive dialogue and progress in policy debates. With facts and evidence taking a backseat to emotional appeals and ideological loyalty, finding common ground becomes increasingly challenging. This polarization limits the potential for bipartisan or multilateral cooperation, essential for addressing complex, cross-cutting issues that require collaborative effort and compromise.

The prioritization of short-term political gains over long-term societal well-being exacerbates these challenges. Politicians and policymakers may be incentivized to pursue policies that offer immediate benefits or appeal to their base, even if such decisions are detrimental in the long run. This short-sighted approach not only undermines the effectiveness of governance but also erodes public trust in institutions and leaders.

Addressing the challenges posed by post-truth politics to governance and policy-making necessitates a comprehensive approach. Restoring the role of facts, evidence, and expert opinion in public discourse is crucial for ensuring that policy decisions are informed and effective. This effort requires bolstering the integrity and credibility of information sources, promoting media literacy among the public, and fostering a political culture that values truth and accountability.

Encouraging open, inclusive dialogue and cooperation across ideological divides can help to bridge the gap between competing perspectives and facilitate the development of consensus-driven policies. By prioritizing long-term societal well-being and embracing the complexities of policy issues, governments and policymakers can navigate the challenges of the post-truth era and work towards more informed, effective governance.

Strengthening Media Literacy and Education

In the battle against the tide of misinformation and the challenges of post-truth politics, strengthening media literacy and enhancing education emerge as crucial strategies. By equipping individuals with the skills to critically evaluate information sources, discern between credible and misleading content, and appreciate the value of evidence-based reasoning, society can foster a more informed and discerning public. Educational initiatives designed to cultivate these skills are essential for empowering citizens to effectively navigate the increasingly complex information landscape that characterizes the digital age.

Media literacy encompasses the ability to access, analyze, evaluate, and create media in a variety of forms. In the context of post-truth politics, media literacy becomes a critical defense mechanism, enabling individuals to sift through the vast amounts of information they encounter daily and identify misinformation. By understanding how media messages are constructed, for what purposes, and by whom, citizens can become more discerning consumers of information and less susceptible to manipulation.

Key components of media literacy education include:

- Source Evaluation: Teaching individuals to assess the credibility of information sources, looking for indicators of reliability such as transparency about sourcing, evidence, and the presence of fact-checking.
- Understanding Biases: Helping individuals recognize both their own biases and those that may be present in media content, including the framing of issues and selective presentation of facts.

- Critical Thinking: Encouraging a questioning attitude towards information and the ability to engage in independent critical thinking, rather than passively accepting media content.
- Digital Literacy: Equipping individuals with the skills to navigate digital media platforms responsibly, including understanding algorithmic content curation and the role of social media in disseminating information.

Educational Initiatives

Effective educational initiatives in media literacy should start from an early age and be integrated into the broader curriculum, ensuring that all students develop a solid foundation in these critical skills. Programs should be adaptive and evolve to address the changing media landscape, incorporating current examples of misinformation and analysis of real-world cases.

Beyond formal education settings, public awareness campaigns and community-based workshops can extend media literacy education to wider audiences, including adults who may not have received such education in school. Partnerships between educational institutions, non-profit organizations, and media outlets can facilitate the development and dissemination of media literacy resources and programs.

The Role of Government and Policy Makers

Government and policy makers play a vital role in supporting media literacy and education initiatives. This can include funding for programs, development of national media literacy strategies, and incorporation of media literacy into educational standards. Additionally, governments can collaborate with technology companies and social media platforms to promote media literacy and reduce the spread of misinformation online.

Strengthening media literacy and education is a vital strategy for empowering citizens to combat misinformation and navigate the complexities of the modern information environment. By investing in educational initiatives that foster critical thinking, source evaluation, and an understanding of media dynamics, society can build a more resilient public, capable of engaging in informed and reasoned discourse. This, in turn, strengthens the foundations of democratic society, ensuring that decisions and debates are informed by evidence and reason rather than misinformation and manipulation.

Rebuilding Trust in Institutions

In the era of post-truth politics, where misinformation and skepticism have eroded public confidence, rebuilding trust in institutions is paramount for the health of democracy. Institutions, from the media to scientific bodies and governmental agencies, play a pivotal role in maintaining the fabric of factual discourse and ensuring the integrity of democratic processes. Restoring this trust necessitates a dynamic approach, focusing on transparency, accountability, active promotion of democratic values, and fostering engagement between the public and experts.

Transparency and accountability are foundational to rebuilding trust. Institutions must commit to openness in their operations, decision-making processes, and the dissemination of information. This includes providing clear, accessible explanations of how decisions are made, the evidence or data supporting those decisions, and the mechanisms in place for public oversight and accountability. By demystifying their processes, institutions can dispel suspicions of bias or hidden agendas, fostering a more trusting relationship with the public.

Institutions should actively communicate their role and importance in safeguarding democratic values and facilitating informed public discourse. This involves not only highlighting their contributions to society but also educating the public on how they operate within the democratic framework. By emphasizing their commitment to facts, evidence-based policy, and the public good, institutions can counter narratives that paint them as self-serving or manipulative.

Encouraging Engagement and Dialogue

Bridging the gap of mistrust also requires direct engagement and dialogue between institutions and the public. Open forums, public hearings, and participatory decision-making processes can foster a sense of inclusion and co-ownership among the public, making them feel more connected to and trusting of institutions. Additionally, these interactions provide opportunities for institutions to directly address public concerns, clarify misconceptions, and demonstrate their responsiveness to public input.

Engagement should also extend to the expert community. Scientists, researchers, and experts should be encouraged to communicate with the public, sharing their knowledge and the relevance of their work to societal well-being. This can help demystify expertise and show the human face of scientific and scholarly endeavors, making expertise more relatable and trusted by the general public.

Reinforcing the Importance of Expertise in Shaping Public Policy

Finally, institutions must champion the role of expertise in public policy. This involves not only consulting with experts in policy-making but also publicly acknowledging the value of expert input and the evidence-based foundations of policy decisions. By doing so, institutions can counteract the narrative that expertise is elitist or disconnected from the public interest, reinforcing the notion that informed, expert-guided decision-making is crucial for addressing complex societal challenges.

Rebuilding trust in institutions is a critical endeavor in the fight against post-truth politics. Through transparency, accountability, active promotion of their democratic role, and engagement with the public and experts, institutions can regain public confidence. This restoration of trust is essential for ensuring that societal debates and policy decisions are grounded in facts and evidence, paving the way for a more informed and cohesive democratic society.

The phenomenon of post-truth politics, characterized by an assault on fact and reason, presents one of the most daunting challenges to contemporary democracy. By undermining the foundations of factual discourse, this trend threatens to unravel the social fabric and impede effective governance. However, through targeted strategies to enhance media literacy, rebuild trust in institutions, and reaffirm the value of evidence-based dialogue, there is hope for reversing the tide of misinformation. This chapter underscores the imperative for collective action in defense of truth and reason, calling upon individuals, media, and governments to champion the cause of informed, rational political engagement for the health of democracy and society at large.

Chapter 11: Surveillance Over Society - The Erosion of Privacy

In the 21st century, the concept of privacy has been profoundly transformed and challenged by the dual forces of technological advancement and evolving societal norms. Once considered a fundamental human right, privacy is now under siege in a world where digital technologies penetrate almost every aspect of our lives. From government agencies collecting vast troves of data to corporations tracking our every online move, the encroachments on personal privacy are pervasive and growing. This chapter seeks to explore the historical context of surveillance, its expansion in the digital era, the consequential impacts on democracy and personal freedom, and the ongoing struggle to balance security needs with the preservation of privacy.

Early Forms of Surveillance

Surveillance, as a concept and practice, is far from a modern invention. Its roots stretch deep into history, with ancient civilizations understanding the value of intelligence gathering and espionage. These early forms of surveillance were crucial in the consolidation of power, the control of vast populations, and the maintenance of empire stability. Records from ancient Egypt, Greece, and Rome, for instance, depict a world where spies were sent out to gather information on enemies and sometimes even on their own citizens.

However, it was in the 20th century that surveillance began to evolve dramatically, marking a departure from these more primitive practices. The Cold War era, in particular, stands out as a period of significant transformation. During these tense times, state surveillance was not just a tool; it became a weapon wielded in the shadows of global politics. Governments on both sides of the ideological divide developed and deployed extensive methods to keep tabs on populations. These efforts were often cloaked in the language of national security and defense against external threats, but they also extended into the control of political thought and the suppression of dissent within borders.

The methods and technologies of surveillance saw rapid advancements during this period. Wiretapping, mail interception, and the use of informants became standard tools in the arsenal of government agencies tasked with internal and external surveillance. These techniques were justified through a mixture of fear, patriotism, and the perceived necessity of maintaining ideological purity and political stability. The impact of these practices on privacy and individual freedoms was profound, setting the stage for the even more complex and pervasive forms of surveillance that the digital age would bring.

This era laid the groundwork for the surveillance state concept, where governments, justified by ever-present security concerns and the need for political control, intrude into personal lives. The legacy of these early forms of surveillance, amplified by technological advancements, continues to influence how modern societies grapple with the balance between privacy and security.

Technological Advancements and Privacy

The dawn of the digital age heralded a seismic shift in the landscape of surveillance, marked by an unprecedented increase in both its capabilities and reach. This transformation was primarily fueled by the transition from analog to digital technologies—a change that significantly enhanced the pervasiveness of surveillance practices while simultaneously making them less conspicuous and, by extension, more insidious.

Central to this shift was the emergence and rapid proliferation of the internet and mobile technologies. The digital footprint left by online activities, combined with the ubiquity of smartphones, provided governments and corporations alike with vast new troves of data to monitor, collect, and analyze. This era also saw the advent of big data analytics, a technological breakthrough that enabled the processing and interpretation of massive datasets at speeds previously unimaginable. Together, these developments have dramatically expanded the scale and scope of surveillance, penetrating deeply into the fabric of everyday life.

In this new digital realm, traditional expectations of privacy and anonymity have been fundamentally altered. Activities that once seemed benign or private, such as online searches, social media interactions, and even location movements tracked by smartphones, have become sources of valuable data for surveillance operations. The ease with which this data can be collected and analyzed means that virtually everyone is subject to some level of surveillance, often without their knowledge or consent.

The implications of these technological advancements are profound. While they have undoubtedly brought significant benefits, such as enhanced communication and access to information, they have also eroded the boundaries of personal privacy. The concept of being unobserved or anonymous has become increasingly elusive, as digital footprints can reveal intimate details about an individual's life, preferences, and behaviors.

This new surveillance paradigm challenges us to reconsider our understanding of privacy in the digital age. It prompts critical questions about the trade-offs between the benefits of technological progress and the preservation of personal freedoms. As surveillance practices continue to evolve in sophistication and reach, the ongoing struggle to balance these competing interests will undoubtedly shape the future of privacy and personal autonomy in the digital era.

The landscape of government surveillance underwent a massive seismic shift in the early 21st century, largely as a response to significant global events that reshaped national security strategies across the world. The terrorist attacks on September 11, 2001, served as a pivotal moment, prompting many governments to significantly expand their surveillance capabilities. This expansion, often framed within the context of preventing future attacks, has led to the implementation of broad and deeply invasive surveillance programs.

In the United States, the passage of the USA PATRIOT Act shortly after the 9/11 attacks marked a dramatic increase in the government's authority to surveil its citizens. This legislation allowed for the collection of telephone metadata, eased restrictions on foreign intelligence gathering within the United States, and expanded the powers of law enforcement agencies to access personal records and conduct wiretaps without traditional legal processes.

The scope of these surveillance activities was further brought to light by Edward Snowden's revelations in 2013. Snowden, a former contractor for the National Security Agency (NSA), leaked classified information detailing the NSA's global surveillance programs. These revelations included programs like PRISM, which collected internet communications from various U.S. internet companies, and the collection of telephone records of millions of Americans by the NSA. Snowden's disclosures sparked a global conversation about privacy, surveillance, and the extent to which governments should be allowed to collect data in the name of national security.

Similar trends have been observed around the world. China, for example, has employed surveillance technology extensively for social control, utilizing a vast network of cameras equipped with facial recognition technology, along with the monitoring of internet activity to maintain and exert control over its population. In Russia and across the European Union, sophisticated surveillance tools have been implemented, ranging from the interception of communications to the use of advanced analytics to monitor and predict potential threats.

These government surveillance programs, often initiated in secrecy and justified under the broad umbrella of national security, have ignited intense debates. At the heart of these discussions is the perennial question of where to draw the line between ensuring security and safeguarding individual freedoms and privacy. As surveillance technologies continue to evolve and become more sophisticated, this debate becomes increasingly complex, challenging societies to find a balance that protects both public safety and the personal liberties that are foundational to democratic governance.

Government Surveillance Programs

The landscape of government surveillance underwent a massive shift in the early 21st century, catalyzed by significant global events that reshaped the discourse on national security and privacy. The terrorist attacks of September 11, 2001, served as a pivotal moment, prompting a worldwide reevaluation of security strategies and the adoption of far-reaching surveillance measures. In the United States, this shift was emblematically represented by the passage of the USA PATRIOT Act. This legislation granted the government unprecedented authority to monitor and collect data on individuals, ostensibly to combat terrorism. Its provisions for wiretapping, data collection, and the scrutiny of financial transactions expanded the surveillance capabilities of agencies like the National Security Agency (NSA) to levels previously unseen.

The revelations by Edward Snowden in 2013 about the NSA's global surveillance programs pulled back the curtain on the scale and depth of these activities. Snowden's disclosures showed not only the breadth of data collection, encompassing phone records, emails, and internet activity, but also the collaboration between the US government and other countries, as well as with major telecommunications and technology companies. This intricate web of surveillance sparked a global debate on privacy, security, and the extent to which governments should be allowed to monitor their citizens and others.

Similarly, in China, the government has implemented an extensive array of surveillance measures that blend technology with social control mechanisms. The use of facial recognition technology, internet monitoring, and the development of the Social Credit System are examples of how surveillance is employed to maintain order and compliance within the society. These practices have raised concerns regarding human rights and the impact on personal freedoms.

In Russia and the European Union, sophisticated surveillance tools have also been deployed, though with varying justifications and outcomes. Russia's use of surveillance technology has been criticized for stifling dissent and controlling public discourse, while in the European Union, efforts to balance security needs with privacy rights have led to stringent data protection laws and debates over the use of surveillance technologies by law enforcement and intelligence agencies.

These government surveillance programs, while often justified under the banner of national security, have ignited intense discussions about the erosion of privacy and individual freedoms. The debate centers on finding a balance between the need for security and the preservation of fundamental rights, a challenge that remains unresolved as technology continues to advance.

Corporate Data Collection and Profiling

In the contemporary digital landscape, surveillance is not solely the domain of government agencies. The corporate sector, particularly technology giants, has emerged as a pivotal player in the gathering and analysis of personal information. Through a myriad of platforms and services that have become integral to daily life, these companies have devised sophisticated mechanisms to compile detailed profiles on millions of individuals globally. This process involves tracking user behaviors, preferences, interactions, and even locations, creating comprehensive digital footprints.

This corporate data collection is often presented under the guise of enhancing user experience—providing free services, tailoring advertisements, and personalizing content. However, this practice brings to the forefront significant ethical concerns regarding privacy and consent. Users, while enjoying the benefits of free platforms and personalized experiences, may not fully understand the extent to which their data is collected, shared, or sold. The transparency around these processes is frequently lacking, leaving users in the dark about the life cycle of their personal information.

The commodification of personal data raises profound questions about the nature of privacy in the digital age. As companies profit from the information gleaned from users' online activities, the line between user benefit and exploitation becomes increasingly blurred. This dynamic underscores a critical tension in the digital economy: the trade-off between the conveniences offered by technological advancements and the intrinsic value of personal privacy.

The implications of corporate data collection extend beyond individual privacy concerns. The aggregation and analysis of vast amounts of data afford these companies unprecedented influence over public opinion, market trends, and even democratic processes. As society grapples with these challenges, the debate continues about the appropriate boundaries for data collection and the role of regulation in protecting individual privacy in a data-driven world.

Erosion of Civil Liberties

The pervasive nature of modern surveillance technologies and methodologies directly undermines the foundation of civil liberties that democratic societies hold dear. Among the most affected freedoms are those of expression, association, and the press—cornerstones of a vibrant and functioning democracy. The widespread knowledge, or even the mere suspicion, that one's actions, communications, and affiliations are constantly monitored exerts a chilling effect on individuals' willingness to speak freely, engage in political activities, or explore controversial ideas. This environment of surveillance-induced self-censorship significantly diminishes the diversity of opinions and ideas circulating in the public sphere, a phenomenon detrimental to the health of public discourse.

The impact of surveillance on civil liberties extends beyond the dampening of free expression. The capacity to monitor, track, and analyze the activities and communications of individuals grants governments—and, by extension, those in power—an unprecedented ability to identify, surveil, and potentially suppress dissenting voices. This capability becomes a potent tool in the hands of authoritarian regimes, where it is often employed to silence opposition, prevent the organization of protests, and clamp down on any form of resistance. Such practices starkly contrast with the democratic ideal of an informed and freely debating society and instead pave the way for the erosion of democratic principles and the consolidation of authoritarian control.

The indiscriminate collection of data and the lack of transparency regarding its use raise serious concerns about the abuse of power and the violation of individual rights. In the absence of robust legal frameworks and oversight mechanisms, the line between legitimate security measures and the infringement on personal freedoms becomes increasingly blurred. As governments and corporations wield these surveillance tools with minimal accountability, the risk of encroachments on privacy and other civil liberties grows, posing a fundamental threat to the principles of democracy and human dignity.

Psychological and Social Consequences

The ramifications of widespread surveillance extend far beyond the realm of politics and law, seeping into the very psyche of individuals and reshaping the social landscape. Living under the perpetual gaze of surveillance technologies can engender deep-seated psychological effects, manifesting as heightened anxiety, a pervasive sense of being scrutinized, and an overarching feeling of powerlessness. This state of constant vigilance not only affects individuals' mental health but also erodes the foundational trust upon which societal institutions and interpersonal relationships are built.

The awareness that one's actions, words, and even thoughts may be monitored and scrutinized dissolves the boundaries between public and private life. This intrusion into what were once considered sanctuaries of personal freedom and autonomy—spaces where individuals could express themselves freely and without judgment—undermines the very essence of individuality. As these private spaces, both physical and digital, become increasingly transparent to outside observers, the opportunity for genuine self-expression and exploration diminishes.

This erosion of privacy and autonomy has far-reaching implications for the social fabric. It stifles the diversity of thought and expression that is critical for the vibrant exchange of ideas and the healthy functioning of society. Moreover, it breeds a culture of conformity and self-censorship, where individuals are reluctant to deviate from the norm or express dissenting opinions for fear of surveillance and its consequences. This not only impoverishes public discourse but also weakens the bonds of community and solidarity, as people grow more suspicious and guarded in their interactions with others.

The psychological toll of surveillance contributes to a fragmented society, where trust is eroded, and social cohesion is undermined. The collective sense of being under surveillance can lead to a breakdown in trust, not only in the institutions responsible for the surveillance but also among individuals themselves. As trust diminishes, the very foundation upon which communities are built—mutual respect, understanding, and cooperation—becomes increasingly unstable. This shift towards a more surveilled and less trusting society poses significant challenges to our ability to connect, empathize, and collaborate with one another, fundamentally altering the nature of social relations in the digital age.

The Justification for Surveillance

In the debate over the pervasive use of surveillance technologies, proponents marshal a range of arguments in support of expanded monitoring capabilities. Central to these arguments is the assertion that surveillance is indispensable for national security, the prevention of terrorism, and the fight against crime. The logic presented is straightforward: in an era where threats can emerge from any quarter and take myriad forms, the ability to monitor communications, track movements, and gather intelligence is vital for the safety and security of society.

Advocates point to instances where surveillance programs have successfully intercepted plans for terrorist attacks, aided in the apprehension of criminals, and contributed to the broader security apparatus tasked with protecting the public. These successes are often showcased as definitive proof of the utility and necessity of surveillance measures, underscoring their role as a linchpin in contemporary security strategies.

However, the invocation of national security and public safety as the primary justifications for surveillance opens a Pandora's box of ethical and philosophical dilemmas. At the heart of this debate is the question of balance: How much privacy are individuals and societies willing to sacrifice for the sake of security? This quandary forces a reckoning with the very values and principles that define democratic societies, challenging us to consider the trade-offs between liberty and security, privacy and protection.

The criteria by which this balance is assessed are often murky and subjective. The effectiveness of surveillance in preventing threats must be weighed against the potential for abuse, the erosion of civil liberties, and the long-term psychological and social impacts. Critics argue that the costs of surveillance—measured in terms of the loss of privacy, the chilling effect on free speech, and the weakening of democratic institutions—may outweigh its purported benefits. They caution against a slide into a surveillance state, where the imperatives of security overshadow fundamental human rights and freedoms.

The justification for surveillance, then, is not a matter of security alone but a complex calculus involving ethical considerations, the preservation of democratic values, and the careful evaluation of the trade-offs that societies are willing to make. As surveillance technologies continue to evolve and permeate further into the fabric of daily life, the debate over their justification and the limits of their use remains a critical conversation for the future of democracy and individual freedom.

Privacy Advocacy and Legal Protections

As the digital age deepens the intrusion into personal privacy, a counter-current has emerged, characterized by a global movement dedicated to the advocacy and defense of privacy rights.

This movement, composed of activists, legal experts, and concerned citizens, has been instrumental in challenging the unregulated expansion of surveillance practices and the commodification of personal data. At the heart of their efforts is a push for robust legal protections that safeguard individual privacy against the overreach of both state and corporate entities.

A landmark achievement in this ongoing battle is the implementation of the General Data Protection Regulation (GDPR) by the European Union. Enacted in 2018, the GDPR represents a comprehensive approach to data protection, setting stringent standards for data collection, processing, and storage. It empowers individuals with unprecedented control over their personal information, including the right to access, correct, and delete their data. The regulation also mandates transparency from organizations regarding data usage and introduces significant penalties for non-compliance, thereby establishing a new benchmark for privacy protections worldwide.

Privacy advocates continue to champion the cause, calling for increased transparency in surveillance practices and the incorporation of judicial oversight to prevent abuses. They argue for the necessity of clear, enforceable boundaries that delineate acceptable from unacceptable forms of data collection and monitoring. Their efforts are aimed at ensuring that surveillance technologies serve the public interest without compromising fundamental freedoms.

The advocacy for privacy rights and the establishment of legal frameworks like the GDPR are crucial steps toward reasserting control over personal information in the digital era. These developments reflect a growing recognition of the need to balance the benefits of technological advancements with the imperatives of privacy and personal freedom. As the landscape of digital surveillance continues to evolve, the dialogue between privacy advocates, policymakers, and technology developers will be pivotal in shaping the future of privacy protections and the preservation of civil liberties in the 21st century.

Technological Solutions and Individual Actions

In the face of increasing surveillance and data collection, individuals are not entirely powerless. A variety of technological solutions and personal practices exist that can significantly enhance privacy protection and reduce the vulnerability of personal information in the digital domain. These measures, ranging from the use of encryption to the adoption of privacy-focused software, provide tangible means for individuals to secure their data and assert greater control over their digital lives.

Encryption: One of the cornerstones of digital privacy, encryption serves as a robust barrier against unauthorized access to data. Utilizing end-to-end encrypted communication tools for messaging and calls ensures that conversations remain confidential, accessible only to the intended recipients.

Secure Communication Tools: Beyond encryption, numerous secure communication apps and platforms prioritize user privacy by implementing features like self-destructing messages and minimizing data retention. Opting for these services can drastically reduce one's exposure to surveillance and data breaches.

Privacy-focused Software and Services: From web browsers and search engines to email providers, a growing ecosystem of privacy-centric solutions offers alternatives to mainstream services that often collect extensive user data. These tools are designed to minimize data collection, not track users across the web, and provide more transparent privacy policies.

Digital Hygiene Practices: Beyond adopting specific tools, cultivating good digital hygiene plays a critical role in protecting privacy. This includes conducting regular privacy settings check-ups on social media and other online accounts, being mindful of the information shared publicly, and thoroughly understanding the terms of service and privacy policies of platforms and apps used.

Educational Resources and Community Support: Engaging with communities focused on digital rights and privacy can provide valuable insights and resources for protecting personal information. Many organizations and forums offer guidance, tutorials, and support for individuals looking to enhance their digital privacy.

By integrating these technological solutions and practices into daily digital routines, individuals can take proactive steps toward safeguarding their personal information. While no single action can provide complete protection in the complex landscape of digital surveillance, a combination of informed choices, technology adoption, and vigilant digital hygiene can significantly bolster one's privacy defenses.

Policy and Societal Changes

Confronting the challenges brought about by pervasive surveillance and the erosion of privacy demands a concerted effort that spans policy reforms and shifts in societal attitudes. The intricate balance between safeguarding privacy and addressing security concerns necessitates a holistic approach, one that reinforces regulatory frameworks, mandates accountability, and cultivates a societal ethos that cherishes privacy as an intrinsic human right.

Strengthening Regulatory Oversight: Central to these efforts is the need for robust regulatory oversight of surveillance technologies and practices. Legislation should be updated to reflect the realities of the digital age, imposing strict guidelines on the collection, use, and sharing of personal data by both governmental bodies and private entities. Such regulations must ensure that surveillance activities are transparent, proportionate, and subject to judicial review, thereby safeguarding against overreach and abuse.

Ensuring Accountability: Accountability mechanisms must be strengthened to ensure that those who deploy surveillance technologies—whether in the public or private sector—are held responsible for their actions. This includes the establishment of independent oversight bodies capable of investigating complaints, auditing surveillance practices, and enforcing penalties for violations of privacy rights.

Fostering a Culture of Privacy: Beyond legal and regulatory measures, fostering a culture that values privacy is crucial. This involves shifting societal attitudes to recognize privacy not as a luxury or an obstacle to security but as a fundamental right that underpins democratic freedoms and personal autonomy. Encouraging a broad societal consensus on the importance of privacy can help resist the normalization of surveillance and the erosion of private spaces.

Public Awareness and Education: Public awareness campaigns and educational initiatives play a pivotal role in this cultural shift. By informing citizens about the risks associated with surveillance, the importance of privacy protections, and the steps individuals can take to safeguard their personal information, these efforts can empower individuals to make informed decisions about their digital lives. Education can also demystify the technical aspects of digital privacy, making it accessible and relevant to a wider audience.

Engagement and Advocacy: Finally, encouraging public engagement and advocacy is vital for driving policy changes and fostering a privacy-conscious society. Activism, public discourse, and participation in the legislative process can influence the development of privacy-friendly policies and promote a more balanced approach to surveillance and security.

Addressing the privacy challenges of the digital age requires a wide-ranging strategy that combines policy innovation, regulatory reform, and a shift in cultural norms. By collectively pursuing these changes, societies can navigate the complexities of surveillance and privacy in the 21st century, ensuring that technological advancements serve to enhance, rather than undermine, individual freedoms and democratic values.

The erosion of privacy in the face of pervasive surveillance represents one of the most pressing challenges of our time, touching upon fundamental human rights, democratic principles, and the very nature of personal freedom. As technology continues to advance, the need for vigilant protection of privacy becomes increasingly imperative. By understanding the complexities of this issue, advocating for robust legal protections, and adopting measures to safeguard personal information, society can navigate the delicate balance between security and privacy. In doing so, we reaffirm the value of privacy as a cornerstone of a free and democratic society, ensuring that individual dignity and autonomy are preserved in the digital age.

Chapter 12: Precarious Futures - The Unraveling Social Contract

The concept of the social contract, which has been instrumental in shaping modern societies, posits that individuals consent, explicitly or implicitly, to surrender some freedoms to a governing body in exchange for protection of their remaining rights and maintenance of the social order. This philosophical cornerstone, central to the development of democratic states and societies, is facing unprecedented challenges in the 21st century. Economic upheaval, technological advancements, political polarization, and environmental crises are testing the resilience and adaptability of the social contract. This chapter seeks to unravel the complex forces at play in the contemporary unraveling of the social contract and to explore pathways towards a reimagined agreement that can address the challenges of our time.

Foundational Theories: The Evolution of the Social Contract Concept

The social contract theory is a monumental framework in the realm of political philosophy, proposing that individuals come together to form a society by consenting, either implicitly or explicitly, to surrender certain freedoms in exchange for the protection of their basic rights and the establishment of order. This concept was not born in a vacuum; it was meticulously developed through the centuries, most notably beginning with Thomas Hobbes and later refined by thinkers such as John Locke and Jean-Jacques Rousseau. Their collective contributions have significantly shaped the principles underlying modern democratic governance, emphasizing the sanctity of individual rights and the pivotal role of consent in the legitimacy of authority.

Thomas Hobbes, through his seminal work "Leviathan," laid the initial groundwork for the social contract theory. Writing in a time of political turmoil in England, Hobbes posited that in the natural state, human life was characterized by a perpetual state of warfare, where existence was "solitary, poor, nasty, brutish, and short." To escape this anarchic condition, he argued, individuals collectively agreed to surrender their absolute freedom to a sovereign or a governing body. This entity, vested with unparalleled authority, was deemed essential for the establishment and maintenance of social peace and order. Hobbes's vision of the social contract was thus fundamentally a trade-off: absolute obedience in exchange for security and peace, highlighting a powerful, albeit somewhat pessimistic, view of human nature and the necessity of a strong central authority.

Building on Hobbes's foundational ideas, John Locke introduced a more optimistic and reciprocal view of the social contract. In his "Two Treatises of Government," Locke argued against the notion of absolute sovereignty and instead proposed a government that derived its legitimacy from the consent of the governed. For Locke, the state of nature was not as grim as Hobbes depicted; it was a state of freedom and equality where individuals had natural rights to life, liberty, and property. The social contract, according to Locke, was formed not to surrender freedom but to establish a governing body whose primary role was to protect these inherent rights. This government was to be limited, with its authority conditional on fulfilling its duty to protect the rights of the individuals, thereby introducing the idea of consent and the right to rebellion against unjust governance.

Jean-Jacques Rousseau further refined the concept of the social contract, emphasizing equality and the collective will of the people. In his work "The Social Contract," Rousseau proposed a model of governance where the general will of the populace was sovereign. Unlike Hobbes and Locke, Rousseau's social contract was not just an agreement to form a society but a continuous process of forming a common will that reflects the collective interests of the people. This general will, in Rousseau's theory, served as the basis for legitimate authority, ensuring that governance remained directly responsive to the needs and desires of the people. Rousseau's contributions shifted the focus towards a more participatory and egalitarian form of governance, highlighting the importance of direct democracy and the principle of equality among citizens.

The philosophical journey from Hobbes to Rousseau illustrates the evolving nature of the social contract theory and its profound impact on the development of democratic governance. By emphasizing the protection of individual rights, the importance of consent, and the ideal of collective will, these foundational theories have laid the groundwork for modern political thought and democratic institutions. The dialogue between the individual and the state, the balance between authority and liberty, and the quest for a just and equitable society continue to be influenced by these seminal ideas, underscoring their enduring relevance in contemporary political discourse.

Adaptations and Transformations: Navigating the Evolving Social Contract

At the intersection of political theory and societal evolution, the social contract stands as a testament to the dynamic relationship between the state and its citizens. Originating from the philosophical inquiries of thinkers like Thomas Hobbes, John Locke, and Jean-Jacques Rousseau, the social contract theory has been the cornerstone of political legitimacy and social organization, outlining the mutual obligations and rights that bind individuals and their governing bodies. However, the steady march of history, marked by seismic shifts in economic, social, and technological landscapes, has prompted a reevaluation of this foundational pact. As societies grapple with the complexities of the 21st century, including global economic fluctuations, technological revolutions, political polarization, and environmental crises, the need for an adapted and transformed social contract has never been more apparent.

The concept of the social contract has been malleable, changing in response to the societal context of its time. The Industrial Revolution was a catalyst for one of the first major transformations, as the shift from agrarian economies to industrialized societies created new economic disparities and social challenges. The response, particularly in Western democracies, was the gradual establishment of welfare states. These states aimed to mitigate the harsher effects of industrial capitalism by providing social safety nets, including public health, education, and social security. This adaptation represented a more collective approach to the social contract, emphasizing the role of the state in ensuring a minimum standard of living for all citizens.

The latter part of the 20th century and the dawn of the 21st have seen the forces of globalization and the ascendancy of neoliberal economic policies challenge and reshape the social contract once again. Globalization has facilitated unprecedented levels of economic interdependence and competition, while neoliberalism has championed deregulation, privatization, and a reduced role for the state in economic management, all in the name of promoting individual responsibility and market freedom. This shift towards neoliberalism has had profound implications for the social contract, often prioritizing market efficiencies and individual entrepreneurship over collective welfare and social solidarity.

As we navigate through the 21st century, the traditional frameworks of the social contract are being tested by diverse and interlinked challenges. Economic upheaval, driven by global market fluctuations and the precarious nature of work in the digital age, raises questions about income security and employment rights. Technological advancements, while offering vast potential for societal improvement, also present concerns around privacy, surveillance, and the displacement of traditional labor. Political polarization and the rise of populist movements reflect growing discontent with established political and economic systems, suggesting a disconnect between citizens and the institutions meant to serve them. Environmental crises, underscored by climate change, resource depletion, and biodiversity loss, demand a reconsideration of our relationship with the natural world and the responsibilities of states and individuals in addressing these existential threats.

The cumulative effect of these challenges is a pressing need to reimagine the social contract. This reimagining involves not only addressing the immediate fissures exposed by economic, social, and environmental crises but also anticipating the future needs and aspirations of societies in an increasingly interconnected and digital world. It requires a holistic approach that balances individual freedoms with collective responsibilities, integrates technological advancements with ethical considerations, and prioritizes sustainable development alongside economic growth.

Modern Interpretations and Applications of the Social Contract

The social contract theory, traditionally rooted in the philosophical musings of Hobbes, Locke, and Rousseau, has undergone substantial transformation in its application and interpretation through the ages, particularly during and after the industrial era. As societies transitioned from agrarian economies to industrial powerhouses, the expectations from and responsibilities of the state also evolved, reflecting a broader understanding of the social contract's scope. This evolution was marked by the integration of social welfare programs, labor rights, and an expanded suite of public services into the contract's framework, signaling a significant shift from the classical focus on security and order to a more nuanced consideration of quality of life and social justice.

The establishment of the welfare state in many countries during the 20th century represented a pivotal moment in the history of the social contract. This expansion was predicated on the acknowledgment that the state's duties extended beyond the mere maintenance of law and order to ensuring a certain standard of living for all its citizens. Social welfare programs, public health initiatives, educational opportunities, and labor rights became integral components of the social contract, embodying the collective commitment to safeguard against the vulnerabilities and inequalities exacerbated by industrial capitalism. This period saw the social contract as a dynamic entity, capable of adapting to the economic and social shifts of the era, and serving as a foundation for more inclusive and equitable societies.

However, the latter part of the 20th century and the early 21st century introduced new dynamics that have tested the resilience and adaptability of the social contract. The wave of globalization, characterized by the unfettered flow of capital, goods, and labor across borders, coupled with the rise of neoliberal economic policies, has brought about a paradigm shift in the interpretation and application of the social contract. Neoliberalism, with its emphasis on deregulation, privatization, and a market-driven approach to economic and social policy, has significantly impacted the welfare state model. This shift towards a more individualistic and market-centric perspective has strained the traditional social guarantees, prompting debates about the adequacy and fairness of the social contract in the face of growing inequalities and social disparities.

The contemporary era, marked by rapid technological advancements, environmental challenges, and increasing social and economic inequalities, calls for a critical reevaluation of the social contract. The question at the heart of this discourse is whether the traditional frameworks of the social contract can withstand the pressures of modern societal demands and global economic forces. This reevaluation involves grappling with the balance between market freedoms and social protections, the role of the state in regulating emerging technologies and industries, and the imperative of integrating environmental sustainability into the core of the social contract. The resilience of the social contract in contemporary society hinges on its ability to adapt to these new realities, ensuring that it remains a relevant and robust foundation for governance, social cohesion, and individual rights in an increasingly complex world.

The journey of the social contract from its philosophical origins to its modern interpretations underscores the theory's enduring relevance and flexibility. As societies continue to evolve, the social contract must also transform, reflecting the changing expectations of citizens and the shifting responsibilities of states. The challenge ahead is to craft a social contract that can navigate the complexities of the 21st century, ensuring that it upholds the principles of equity, justice, and mutual respect in an era of unprecedented change.

The Impact of Globalization on the Social Contract

Globalization, characterized by the increased interconnectedness and interdependence of the world's economies, societies, and cultures, has been a driving force of change in the 21st century. This phenomenon has transcended the traditional boundaries that once defined national economies and social contracts, ushering in an era of global economic integration that has fundamentally altered the landscape within which states and their citizens interact. While globalization has undeniably fostered economic growth, expanded global trade, and facilitated unprecedented levels of investment, it has also presented significant challenges to the social contract, particularly in terms of job security, labor rights, and income distribution.

The essence of globalization lies in its ability to erode the traditional boundaries that have historically delineated national markets and political sovereignties. This erosion has not only facilitated the free flow of capital, goods, services, and labor across borders but has also introduced a complex web of economic interactions that defy conventional governance structures. As a result, the national social contract, which relies on the ability of states to regulate economic activities and protect the welfare of their citizens, faces new challenges in asserting its relevance and effectiveness in a globalized context.

The economic integration fostered by globalization has yielded a mixed bag of outcomes. On one hand, it has contributed to economic efficiency, innovation, and the lowering of consumer prices through competitive markets. On the other hand, this integration has led to significant job displacement, as industries move to regions where labor and production costs are lower. The phenomenon of offshoring and outsourcing has not only affected manufacturing jobs but has also started to impact the service sector, leaving behind a trail of economic uncertainty and job insecurity for many workers.

The weakening of labor protections can be directly attributed to the competitive pressures of a globalized economy. As countries vie for foreign investment and industrial projects, there is a tendency to dilute labor laws and protections to attract multinational corporations. This race to the bottom not only undermines workers' rights but also challenges the very foundation of the social contract that promises security and welfare for all citizens.

Perhaps one of the most concerning impacts of globalization is the exacerbation of income inequality. The benefits of globalization have not been distributed evenly across or within societies, leading to a widening gap between the wealthy and the poor. The consolidation of wealth and power in the hands of a global elite, coupled with the marginalization of lower and middle-income workers, poses a significant threat to social cohesion and the stability of democratic institutions. This growing inequality challenges the state's ability to fulfill its obligations under the social contract, as it struggles to provide equitable opportunities and protections for all citizens.

Globalization has undoubtedly reshaped the parameters within which the social contract operates, presenting both opportunities and challenges. While it has driven economic growth and global connectivity, it has also exposed vulnerabilities in national governance models, labor markets, and social welfare systems. The impact of globalization necessitates a reimagining of the social contract, one that can navigate the complexities of a globalized world while ensuring the protection and prosperity of all citizens. Addressing these challenges requires innovative policy solutions, international cooperation, and a commitment to preserving the fundamental principles of equity and social justice that underpin the social contract.

The Gig Economy and Precarious Work: Eroding the Foundations of the Social Contract

In recent years, the labor market has experienced significant transformations, most notably through the rise of the gig economy and the increasing automation of labor. This shift marks a move away from traditional, stable employment models towards more precarious forms of work. The gig economy, characterized by short-term contracts or freelance work as opposed to permanent jobs, alongside the rapid advancement of technology leading to automation, has fundamentally altered the landscape of employment. These changes pose profound challenges to the established social contract, particularly in terms of employment protections and benefits that workers have historically relied upon.

The traditional social contract, underpinned by stable, full-time employment, offered workers a sense of security through various protections and benefits, including healthcare, pensions, and unemployment insurance. However, the nature of gig work—marked by its temporary, flexible, and often independent contractor status—significantly undermines these protections. Gig workers, encompassing a wide range of activities from ride-sharing to freelance programming, often find themselves without the safeguards afforded to traditional employees. This lack of security and benefits leaves them exposed to greater economic instability and vulnerability.

The emergence of precarious work highlights the inadequacies of existing social safety nets designed for a bygone era of employment. The gig economy and automation-driven job displacement challenge the very premise of the social contract's promise: security in exchange for labor. As work becomes more fragmented and uncertain, the mechanisms in place to support workers during periods of unemployment, illness, or retirement are revealed to be ill-equipped to handle the nuances of modern employment practices. This gap between the changing nature of work and the stagnant evolution of social protections raises urgent questions about how to adapt and reimagine social safety nets for the 21st century.

The rise of precarious work demands a rethinking of the social contract to ensure it remains relevant and effective in protecting workers in the modern economy. This may involve exploring new models of social security that account for the variability of gig work, such as portable benefits that follow workers across jobs and sectors. Additionally, there's a need for policies that address the impact of automation on employment, potentially through retraining programs, universal basic income schemes, or other innovative solutions aimed at supporting individuals through the transition.

The gig economy and the automation of labor represent not just shifts in the mode of employment but signal a deeper erosion of the social contract as we know it. The movement towards more precarious forms of work challenges the foundational promise of security traditionally offered in exchange for labor, calling for a significant reevaluation and adaptation of social safety nets. As we navigate these changes, the imperative to reshape the social contract to reflect the realities of the modern workforce becomes increasingly clear, ensuring that all workers are afforded dignity, security, and protection in the face of evolving employment landscapes.

Digital Surveillance and Privacy: A Modern Challenge to the Social Contract

The advent of the digital age has brought with it transformative technologies that, while facilitating global connectivity and access to information, have also significantly expanded the capacity for surveillance. Today, both government and corporate entities wield the power to monitor, collect, and analyze vast quantities of personal data. This capability to surveil has raised pressing concerns regarding privacy and freedom of expression, two fundamental provisions traditionally protected by the social contract. As these entities look deeper into the private lives of individuals, often without explicit consent or sufficient oversight, a critical question emerges: What rights do individuals retain in a society increasingly characterized by omnipresent surveillance?

Government surveillance programs, often justified on the grounds of national security, pose intricate challenges to the social contract. In the pursuit of safeguarding citizens from threats, governments have increasingly adopted surveillance technologies that can intrude into personal privacy. This tension between the need for security and the right to privacy forces a reevaluation of the social contract, questioning the balance of power between the state and its citizens. The absence of adequate oversight mechanisms and transparency in these surveillance practices further complicates the issue, leading to an environment where citizens' rights and freedoms can be compromised under the guise of security.

Parallel to government surveillance, the digital age has seen the rise of corporate surveillance, with businesses collecting and analyzing personal data on an unprecedented scale. This data is often used for targeted advertising, market research, and the personalization of services, transforming personal privacy into a commodity to be traded. Unlike government surveillance, corporate surveillance operates in a predominantly unregulated space, raising significant concerns about consent and control over personal information. The pervasive nature of digital tracking and data collection challenges the traditional boundaries of the social contract, prompting questions about autonomy, consent, and the commodification of personal life.

The challenges posed by digital surveillance necessitate a reimagining of the social contract to ensure it adequately protects individuals' rights in the digital age. This entails establishing clear norms and regulations that govern the collection, use, and sharing of personal data by both government and corporate entities. Ensuring transparency, enhancing oversight, and securing meaningful consent are pivotal to safeguarding privacy and freedom of expression. Moreover, fostering digital literacy among citizens is crucial in empowering individuals to navigate the complexities of digital surveillance and protect their personal autonomy.

As surveillance capabilities continue to grow in scope and sophistication, the imperative to adapt the social contract to the realities of the digital age becomes increasingly urgent. Balancing the benefits of technological advancements with the protection of fundamental rights requires a collaborative effort among governments, corporations, and civil society. By redefining the parameters of privacy and freedom in the digital realm, society can work towards a renewed social contract that upholds the dignity and autonomy of individuals in an increasingly surveilled world.

Artificial Intelligence and Employment: Navigating the Future of Work

The advent of Artificial Intelligence (AI) and robotics in the workplace marks a pivotal moment in the evolution of employment and, consequently, the social contract. These technological innovations offer the potential to significantly enhance productivity and create new opportunities for economic growth. However, they also pose substantial challenges, particularly in the form of job displacement and the potential obsolescence of certain types of work. As AI systems become more sophisticated, capable of automating not just manual labor but cognitive tasks as well, the dual nature of these technologies—as both enablers and disruptors—becomes increasingly apparent.

On the one hand, AI and robotics can drive the creation of new jobs and industries, much as previous technological advancements have. These fields demand new skills and expertise, from AI system development and maintenance to data analysis and cybersecurity, highlighting the potential for significant productivity gains. Moreover, AI can augment human labor, freeing workers from repetitive tasks to focus on more complex, creative, or interpersonal activities. This synergy between human and machine labor could unlock unprecedented levels of efficiency and innovation within the workplace.

On the other hand, the rapid advancement and adoption of AI pose a threat to traditional employment models. Many jobs, especially those involving routine, predictable tasks, are at high risk of automation. This shift threatens not just individual positions but entire sectors, with the potential to disrupt labor markets and exacerbate income inequality. The specter of mass unemployment or underemployment raises pressing questions about the distribution of wealth generated by AI-driven productivity gains and the mechanisms in place to support those displaced by technology.

The implications of AI and robotics for employment necessitate a profound rethinking of the social contract, especially concerning work, income distribution, and social welfare. Traditional models, which tie social benefits and security closely to employment, may no longer be adequate in a landscape where stable, long-term jobs are less common. This situation calls for innovative policy solutions, such as universal basic income (UBI), retraining and reskilling programs, and new forms of worker organization and representation, to ensure that the benefits of AI are broadly shared and that individuals are protected against the risks of rapid technological change.

Artificial Intelligence and robotics represent a frontier of both opportunity and challenge for the labor market and the social contract. As these technologies continue to evolve and reshape the nature of work, the need to adapt our social, economic, and policy frameworks becomes increasingly urgent. Ensuring that AI serves to enhance, rather than undermine, the social fabric will require concerted effort from policymakers, businesses, and individuals alike. By embracing innovation while safeguarding against its potential drawbacks, we can navigate the transition to an AI-driven economy in a way that promotes inclusivity, equity, and human well-being.

Climate Change and Social Equity: Reframing the Social Contract for Ecological Stewardship

The escalating crisis of climate change presents not just an environmental challenge, but a profound ethical and social dilemma, necessitating a critical reevaluation of the traditional social contract. The devastating impacts of climate change, including extreme weather events, rising sea levels, and shifting climate patterns, underscore the urgency of integrating ecological stewardship directly into the foundational agreements that govern societal relationships and responsibilities. This integration is crucial not only for the preservation of the planet but also for ensuring social equity and justice, as the burdens of environmental degradation disproportionately fall on marginalized communities and future generations.

Climate change amplifies existing social and economic inequalities, hitting hardest those who are least responsible for greenhouse gas emissions. Marginalized communities, often lacking the resources to adapt to environmental changes, face greater risks from pollution, food scarcity, and displacement. Moreover, the ethical consideration of future generations, who will bear the brunt of today's environmental decisions, introduces an intergenerational dimension to social equity. These disproportionate impacts highlight the inadequacies of a social contract that neglects the environmental determinants of social welfare and justice.

Addressing the challenges posed by climate change requires a fundamental shift in how societies conceptualize and enact the social contract. This entails embedding sustainability and ecological responsibility at the core of social agreements, ensuring that economic development and social policies do not come at the expense of environmental health. Integrating sustainability principles involves rethinking economic indicators to value environmental preservation and social well-being, incentivizing green technologies, and promoting sustainable practices that safeguard the planet for current and future inhabitants.

The integration of climate change and social equity considerations into the social contract has significant policy implications. It calls for robust climate action that includes stringent emissions reductions, investment in renewable energy, and the protection of natural ecosystems. Additionally, social policies must be designed to ensure that the transition to a green economy is just and inclusive, providing support for communities and workers affected by the shift away from fossil fuels. This might include job retraining programs, equitable access to clean technologies, and mechanisms for community participation in environmental decision-making.

The interconnections between climate change, social equity, and the social contract underline the necessity of a holistic approach to addressing the environmental crisis—one that considers the ethical, social, and economic dimensions of sustainability. By reimagining the social contract to embrace ecological stewardship and social justice, societies can forge a path towards a sustainable future that honors the rights and dignity of all beings, present and future. The task ahead is formidable, but the imperative for action has never been clearer, as the well-being of the planet and the equity of its inhabitants hang in the balance.

Sustainable Development and Inter-generational Equity: Reimagining the Social Contract

As the world grapples with the escalating crisis of climate change and environmental degradation, it becomes increasingly clear that our existing social frameworks must evolve. The concept of the social contract, traditionally focused on the relationship between individuals and the state, now faces a pressing need for expansion to include the principles of sustainable development and inter-generational equity. This shift represents not merely an adaptation but a fundamental reconceptualization of the social contract to ensure the health and viability of our planet for both present and future generations.

Sustainable development, defined as meeting the needs of the present without compromising the ability of future generations to meet their own needs, must become a cornerstone of the social contract. This approach necessitates a holistic view of economic growth, social inclusion, and environmental protection as interconnected pillars that support the collective well-being of humanity. Incorporating sustainable development into the social contract involves a commitment to green technologies, renewable energy sources, and conservation practices that preserve the Earth's natural resources. It also means designing economic and social policies that promote equity and address the root causes of poverty and inequality.

The concept of inter-generational equity is fundamental to this new vision of the social contract. It emphasizes the moral and ethical obligation to pass on a world that is as good as, if not better than, the one inherited. This principle challenges short-term thinking and planning, advocating for policies and actions that consider their long-term impact on the planet and future inhabitants. Addressing climate change, preserving biodiversity, and reducing environmental degradation are critical components of ensuring inter-generational equity.

Embedding sustainable development and inter-generational equity into the social contract requires transformative changes at both national and global levels. Governments, corporations, and civil society must work together to enact policies that promote sustainable practices and equitable growth. This includes investing in green infrastructure, transitioning to sustainable agriculture, and implementing fair trade practices. Additionally, global cooperation is essential to address transboundary environmental issues, such as climate change and ocean pollution, which require coordinated efforts beyond individual national capacities.

The environmental challenges of our time demand a reimagined social contract that places sustainable development and inter-generational equity at its core. This reconceptualization is not just about adapting to environmental crises but about creating a framework for living that respects the planet's ecological boundaries and ensures fairness and justice for all its inhabitants, now and in the future. By committing to sustainable development and inter-generational equity, we can forge a social contract that supports a thriving, equitable, and sustainable world for generations to come.

 Inclusive Governance and Participation: Foundations of a Renewed Social Contract

In the complex web of modern societies, the principles of inclusive governance and broad participation stand as essential pillars for a renewed social contract. As we navigate through an era marked by rapid technological advances, social changes, and increasing globalization, the imperative for a governance model that embraces diversity, ensures equity, and fosters democratic engagement has never been more critical. This approach to governance recognizes the value of every citizen's voice and the importance of reflecting the needs and aspirations of the entire populace in policy-making and democratic processes.

At the heart of inclusive governance is the commitment to transparency and accountability. These principles act as the bedrock for building trust between the government and its citizens, ensuring that decision-making processes are open, understandable, and subject to scrutiny. Enhancing transparency involves making information about governmental policies, plans, and actions readily available and accessible to all. Accountability requires that officials are answerable for their actions and decisions, facing appropriate consequences when failing to meet the established standards of integrity and efficacy. Together, transparency and accountability create a governance environment where corruption is minimized, and public confidence is maximized.

Broadening participation in democratic processes is another critical aspect of forging a more resilient and equitable social contract. This involves creating ample opportunities for citizens to engage in public discourse, decision-making, and policy formulation. It means going beyond traditional voting mechanisms to include public consultations, participatory budgeting, and the use of digital platforms to gather input and feedback from the wider community. By actively involving citizens in governance, societies can ensure that diverse perspectives are considered, marginalized voices are heard, and policies more accurately reflect the collective will and interests of the populace.

While the benefits of inclusive governance and participation are clear, implementing these principles presents its own set of challenges. Ensuring inclusivity requires deliberate efforts to dismantle barriers to participation, such as economic disparities, lack of education, and social prejudices. It also involves adapting to the digital age, where information technology can both facilitate and hinder democratic engagement. Addressing these challenges necessitates innovative solutions, commitment from all societal sectors, and a continuous process of learning and adaptation.

A renewed social contract rooted in inclusive governance and broad participation is essential for building societies that are resilient, equitable, and reflective of the diverse needs of their citizens. By prioritizing transparency, accountability, and public engagement, we can create a democratic framework that empowers every individual, bridges divides, and harnesses the collective strength of the community. In doing so, we lay the groundwork for a social contract that not only responds to the challenges of the modern era but also anticipates and shapes a brighter, more inclusive future.

Universal Basic Income and Social Safety Nets: Reinforcing the Social Contract

In an era marked by rapid technological advancement, economic volatility, and the growing precarity of work, the social contract that has underpinned modern societies finds itself at a pivotal juncture. The promise of security and prosperity in exchange for participation in the social and economic life of a nation is increasingly under strain. In response to these challenges, innovative policy solutions such as Universal Basic Income (UBI) emerge as potential mechanisms to bolster the social contract, ensuring that it remains relevant and robust in addressing the needs of all citizens in a changing world.

Universal Basic Income represents a radical rethinking of social welfare, proposing a guaranteed, unconditional income provided by the state to every citizen, irrespective of employment status or income level. This bold initiative aims to address the growing uncertainties of modern employment, characterized by gig economy jobs, automation, and fluctuating labor markets, by providing a stable financial foundation for all individuals. UBI is envisioned not just as a safety net, but as a cornerstone of a new social contract that ensures economic security and personal dignity in the face of pervasive changes.

The introduction of UBI could significantly mitigate the precarity associated with contemporary employment landscapes. By guaranteeing a basic income, individuals are afforded a degree of financial stability, enabling them to navigate periods of unemployment, transition between jobs, or pursue education and training without the immediate pressure of financial hardship. Furthermore, UBI empowers individuals to make choices about their work and life, potentially fostering greater creativity, entrepreneurship, and participation in community and civic activities, thus enriching the social fabric.

In addition to UBI, the enhancement of social safety nets is crucial in adapting the social contract to current realities. This entails not only financial support but also access to healthcare, education, and housing, ensuring a comprehensive approach to social welfare that addresses the current nature of human security. By fortifying social safety nets, societies can create more resilient and inclusive systems that protect and empower their most vulnerable citizens, reflecting a collective commitment to welfare and equity.

The consideration of Universal Basic Income and the strengthening of social safety nets represent crucial steps in reimagining the social contract for the 21st century. In the face of economic, technological, and environmental pressures, the need for a renewed social contract that embraces inclusivity, sustainability, and adaptability is more pressing than ever. By prioritizing innovative solutions like UBI, alongside a commitment to broadening participation in democratic processes and enhancing social equity, societies can forge a new social contract that upholds the principles of equity, justice, and security for all. This chapter has charted a path toward such a transformation, envisioning a future where the social contract continues to serve as a foundational element of society, adaptable and robust in the face of the complexities and uncertainties of modern life.

Part IV: Cultural and Ethical Reflections

Chapter 13: The Culture of Excess - Consumerism Unchecked

In contemporary society, consumerism stands as a defining characteristic, influencing not only economic activities but also cultural norms and personal identities. Driven by a relentless pursuit of material possessions, consumer culture has permeated every aspect of life, from the way we define success and happiness to how we engage with the world around us. This chapter aims to dissect the phenomenon of consumerism, tracing its historical roots, examining its psychological and social dimensions, and confronting the environmental and economic challenges it presents. Through this exploration, we will look into the ethical considerations raised by a culture of excess and consider pathways toward a more sustainable and conscientious approach to consumption.

From Scarcity to Abundance

The journey from societies where daily life revolved around subsistence living to ones overwhelmed by the forces of mass production and consumption is one of the most remarkable transitions in human history. At the heart of this transformation was the Industrial Revolution, a period of great technological innovation and industrial growth that began in the late 18th century. This era introduced machinery and manufacturing processes that dramatically increased the production of goods, altering the very fabric of society. Before this pivotal change, most communities were defined by scarcity; goods were handmade, labor-intensive, and produced in relatively small quantities. This meant that for the average person, life was about meeting basic needs, with little room for the concept of excess or luxury.

The post-World War II era marked another significant leap forward for consumer culture, especially in the Western world. The devastation of the war had ironically laid the groundwork for economic prosperity in many parts of the globe. Rebuilding efforts led to innovations in manufacturing and a boom in construction, while technological advancements in communication and transportation further shrunk the world, making goods and services more accessible than ever before. During this time, the concept of "consumer" took on new weight and significance. Economic activity began to center increasingly around consumption rather than production, with success and social progress measured by the ability to consume.

This era of abundance was characterized by the rapid expansion of the middle class, which now had access to a level of wealth and leisure previously unimaginable. The availability of credit expanded, making it possible for more people to purchase homes, automobiles, and household appliances, thereby fueling the economy even further. Advertising became a powerful tool, shaping desires and creating a culture where identity and social status were intricately linked to one's possessions. For the first time in history, the majority of people in industrialized nations could afford to buy more than what was strictly necessary for survival, heralding a new age where consumerism was not just a way of life but a marker of civilization's progress.

As we moved further into the 20th and 21st centuries, the culture of consumption has continued to evolve, with digital technology and globalization introducing new ways of accessing goods and services. Yet, the transition from scarcity to abundance has not been without its challenges. The environmental, economic, and social implications of unchecked consumerism are becoming increasingly apparent, prompting a reevaluation of what true progress means in the context of sustainable and equitable development. The story of this transformation is a testament to human ingenuity and the desire for improvement, but it also serves as a cautionary tale about the limits of consumption-driven growth.

The Role of Advertising and Media

      Advertising and mass media have played a pivotal role in the exponential growth of consumer culture. Beyond the simple promotion of products, they have woven complex narratives that intertwine ideals of beauty, success, and happiness with the acquisition of material possessions. Through sophisticated marketing strategies and the construction of compelling brand identities, companies have not just sold items; they've sold lifestyles and dreams.

Lifestyle marketing has emerged as a particularly influential force, effectively blurring the distinctions between products and personal identity. This approach goes beyond traditional advertising methods, which might focus on the practical benefits or features of a product. Instead, it taps into the emotional and aspirational desires of the consumer, suggesting that purchasing a certain product is not merely a transaction but an entry point into a desired way of life. Brands position themselves as gatekeepers to realms of exclusivity, achievement, or fulfillment, making the act of consumption feel integral to personal development and social acceptance.

This manipulation of consumer desires has been facilitated by the global reach and persuasive power of mass media. Television, magazines, and, more recently, social media platforms, have allowed advertisers to embed their messages into the daily lives of billions of people. Through repeated exposure to idealized images and narratives that link happiness and success with consumption, individuals are encouraged to pursue an endless cycle of buying and discarding, always in search of the next product that promises to bring them closer to the lifestyle portrayed.

The advent of digital marketing has introduced new levels of personalization and targeting, making it possible for advertisers to tailor their messages to the specific desires and insecurities of individual consumers. The feedback loop created by online behavior tracking further refines these strategies, creating a cycle where consumption is not just a response to advertising but a core component of identity and social belonging.

This deep embedding of consumerism within social and cultural norms has profound implications. It shapes how individuals perceive themselves and others, how they define success and happiness, and even how they interact with the world around them. As consumer culture continues to evolve with advancements in technology and media, the role of advertising in shaping societal values and personal aspirations remains both significant and contentious. The challenge lies in navigating the fine line between inspiration and manipulation, ensuring that the pursuit of material goods does not overshadow more sustainable and equitable paths to fulfillment and well-being.

The Pursuit of Happiness Through Consumption

At the core of consumerism's appeal is the powerful and pervasive notion that acquiring material possessions is synonymous with achieving happiness and fulfillment. This concept has been deeply embedded in the collective consciousness, continually reinforced by relentless advertising and prevailing cultural narratives that equate success with the accumulation of wealth and objects. The cycle of desire and acquisition that this belief system fosters has become a defining feature of consumer behavior, shaping not only individual choices but the very values that societies uphold.

The process is seemingly straightforward and immensely appealing: the acquisition of a new product leads to a temporary surge of joy, reinforcing the belief that happiness can be bought. This cycle of consumption is self-perpetuating, driven by the continuous creation of new desires and the promise of satisfaction through purchase. As a result, consumer culture promotes a constant pursuit of the "next big thing," with individuals often believing that the next purchase will bring the contentment that previous ones did not.

However, a growing body of research challenges this fundamental assumption of consumerism. Studies in psychology and sociology consistently indicate that the happiness derived from material acquisition is fleeting at best. While new possessions can provide a temporary boost in mood and self-esteem, they rarely contribute to long-term satisfaction or well-being. Instead, this relentless pursuit can lead to a sense of emptiness, a void that cannot be filled by material means alone.

The emphasis on material wealth and possessions as the primary indicators of success and well-being has also been linked to negative outcomes such as increased anxiety, depression, and a sense of social isolation. The comparison trap, fueled by social media and advertising, exacerbates these feelings, as individuals measure their own worth and achievements against the curated and often unrealistic portrayals of others' lives.

Critically, the realization is dawning that happiness and fulfillment are far more complex and diverse than consumer culture suggests. Factors such as meaningful relationships, a sense of community, personal growth, and contributions to society play a crucial role in achieving lasting satisfaction. This growing awareness is prompting a reevaluation of values and a shift towards more sustainable and fulfilling ways of living, challenging the consumerist model and its narrow definition of happiness.

Social Stratification and Consumer Identity

Within modern society, consumer goods transcend their mere functional purposes, evolving into potent symbols of social status, identity, and belonging. The fervent pursuit of branded merchandise, luxury commodities, and cutting-edge technological gadgets is frequently motivated by an individual's desire to project signals of wealth, discerning taste, and a sense of belonging to a particular social echelon. This dynamic plays a pivotal role in reinforcing and exacerbating existing structures of social stratification, effectively using access to, and possession of, certain products as a barometer for determining an individual's standing within the societal hierarchy.

The implications of this phenomenon extend beyond the realm of personal consumption choices, deeply influencing the fabric of social interactions and the construction of personal identity. In a culture increasingly dominated by consumerism, the possession of certain items becomes a key determinant of social value, with individuals often judged based on the brands they wear, the gadgets they use, and the lifestyle they can afford. This creates a competitive environment where one's self-worth and social recognition are contingent upon one's ability to consume and display symbols of economic capital.

This culture of comparison and competition fosters an environment where social relations are commodified, and personal value is measured by material success. The relentless drive to accumulate symbols of status not only exacerbates social divisions but also promotes a superficial understanding of identity and community. In such a landscape, the quest for social belonging often leads to a paradoxical sense of isolation, as relationships become transactional and contingent on the ability to participate in the consumerist cycle.

The deep-seated association between consumer goods and social identity also underscores the challenges of breaking free from the cycles of consumption and competition. As individuals navigate their social worlds, the pressure to conform to prevailing norms of consumption can be overwhelming, leading to a perpetual cycle of acquisition and display. This dynamic highlights the need for a critical examination of the values that underpin consumer culture and its impact on social cohesion, individual well-being, and the broader fabric of society.

The Environmental Toll

The relentless drive of consumerism has led to significant environmental consequences, marking a troubling aspect of modern society's relationship with the natural world. As consumer demand continues to grow, the strain on Earth's resources intensifies, highlighting the unsustainable nature of current consumption patterns. The environmental toll of such practices is both profound and far-reaching, encompassing the depletion of natural resources, escalating pollution levels, and the generation of an overwhelming amount of waste.

One of the most illustrative examples of these environmental impacts is the fast fashion industry. Characterized by rapid production cycles, low-cost materials, and a constant churn of new styles, fast fashion epitomizes the ethos of disposable consumerism. This industry alone is responsible for significant water consumption, the use of toxic chemicals in production processes, and the creation of substantial textile waste. Garments are produced in massive quantities, sold at low prices, and often discarded after a few wears, contributing to the growing problem of landfill waste.

The environmental degradation associated with consumerism extends beyond the fashion industry, affecting every sector of the economy. The extraction and consumption of fossil fuels for energy, the exploitation of minerals for electronics and other goods, and the unsustainable agricultural practices for food production all contribute to a global ecological crisis. These activities not only exhaust finite resources but also lead to habitat destruction, biodiversity loss, and significant contributions to climate change.

The realization of the environmental cost of unchecked consumerism has prompted a critical reevaluation of how goods are produced and consumed. There is a growing recognition of the need for more sustainable practices that prioritize long-term ecological health over short-term consumer satisfaction. This includes adopting circular economy principles, where products are designed for longevity, reuse, and recyclability, reducing the demand on raw materials and minimizing waste. It also involves a shift in consumer behavior towards more mindful and sustainable consumption choices, recognizing that the health of our planet is intrinsically linked to the ways in which we choose to live and consume.

Economic Disparities and Global Inequities

The pervasive reach of consumerism extends into the economic realm, where its effects contribute to widening the gap between the wealthy and the poor, both within nations and globally. The modern consumer economy, characterized by its globalized supply chains, has deepened economic disparities and fostered a cycle of exploitation and inequality that undermines the principles of fairness and sustainability.

At the heart of this issue is the reality that the global economy relies heavily on labor from developing countries to produce goods for the consumer markets of more affluent nations. These supply chains often operate in regions where labor laws are less stringent or less enforced, allowing for conditions that can be exploitative. Workers in these settings are frequently subjected to poor working conditions, long hours, and wages that do not meet basic living standards, all in service of producing goods at the low costs demanded by consumers and corporations in wealthier countries.

This system of global trade and production not only perpetuates economic disparities but also raises significant ethical questions about the fairness of consuming goods produced under such conditions. The pursuit of lower production costs and higher profits by multinational corporations often comes at the expense of workers' rights and well-being in less developed economies. Furthermore, the environmental degradation associated with these production practices disproportionately affects the poorest and most vulnerable populations, who are least equipped to cope with the impacts of pollution and climate change.

The reliance on exploitative labor practices and the resultant economic inequities highlight the need for a reevaluation of global consumption patterns. Advocates for change argue for the implementation of fair trade practices, stricter regulation of international labor standards, and greater corporate accountability. By addressing the root causes of economic disparities and working towards more equitable and sustainable models of production and consumption, there is potential to create a more just global economy that benefits all participants.

The conversation around consumerism and its impacts is evolving to include a broader understanding of the interconnectedness of environmental, social, and economic issues. As awareness grows, there is increasing pressure on individuals, corporations, and governments to consider the long-term implications of their consumption choices and to work towards a more equitable and sustainable future.

The Morality of Consumption

In the face of a culture steeped in excess, the ethical implications of our consumption habits emerge as a pivotal concern. The rampant consumerism characterizing much of modern society invites a critical examination of the moral obligations that individuals, corporations, and governments hold in addressing and alleviating its detrimental effects. This contemplation extends beyond mere critique, urging a collective shift towards consumption practices that are mindful of their broader impact on environmental sustainability, social justice, and economic fairness.

The environmental dimension of consumer ethics challenges us to steward the planet's resources responsibly. As the effects of climate change become increasingly undeniable, the moral imperative to reduce waste, curb pollution, and shift towards renewable energy sources becomes paramount. This responsibility not only lies with individual consumers in their daily choices but also with corporations in their production methods and governments in their policies and regulations.

Social justice, intertwined with ethical consumption, calls for a reflection on how our buying habits affect communities around the globe. It raises questions about the fairness of labor practices, the exploitation of vulnerable populations, and the role of consumer demand in perpetuating inequalities. Ethical consumption in this context advocates for supporting fair trade, demanding transparency in supply chains, and choosing products that contribute to the well-being of workers and communities.

Economic equity, as a component of the morality of consumption, prompts consideration of how consumerism influences wealth distribution and access to resources. The pursuit of material wealth often exacerbates economic disparities, highlighting the need for a more equitable system that ensures access to basic needs and opportunities for all. Ethical approaches to consumption recognize the importance of supporting economies that favor sustainability and equity over unchecked growth and profit maximization.

Confronting the morality of consumption requires an approach that considers the complex interplay of environmental, social, and economic factors. It demands of us a willingness to reevaluate our values and make conscientious choices that align with a vision of long-term well-being and justice. By embracing a more ethical paradigm of consumption, we can contribute to a more sustainable, equitable, and compassionate world.

Towards Sustainable and Ethical Consumption

In the quest to mitigate the adverse effects of consumerism, a pivotal shift towards sustainable and ethical consumption practices is paramount. This transition involves embracing innovative models and ideologies that promise to lessen our ecological footprint and promote fairness within global markets. Among these, the circular economy emerges as a compelling framework, advocating for the reuse, recycling, and efficient utilization of resources to minimize waste and reduce environmental degradation. By designing products with their entire lifecycle in mind, this model encourages a departure from the traditional, linear approach of "take-make-dispose," aiming instead for a regenerative cycle that sustains resource use within ecological limits.

Parallel to the circular economy, movements like minimalism and ethical consumerism offer powerful counter-narratives to the prevailing culture of excess. Minimalism, with its emphasis on simplicity and intentionality in consumption, challenges the notion that happiness and success are predicated on the accumulation of material possessions. It invites individuals to reconsider what they truly need for a fulfilling life, advocating for a focus on experiences and relationships over goods and gadgets. Ethical consumerism, on the other hand, underscores the importance of transparency, fairness, and sustainability in the production and distribution of goods. It encourages consumers to make informed choices that support ethical business practices, fair labor conditions, and environmental stewardship.

By aligning with these principles, consumers wield the power to drive meaningful change, advocating for a world where economic activities contribute positively to social equity and environmental health. This shift requires not only individual action but also collective efforts to redefine societal norms and values around consumption. It calls for a critical examination of our consumption habits and a concerted move towards practices that prioritize long-term well-being over immediate gratification.

The challenges of unchecked consumerism—encompassing environmental degradation, social inequality, and economic instability—demand a comprehensive and informed response. As this chapter has delineated, understanding the nuanced dynamics of consumerism is the first step towards cultivating a culture that values sustainability, mindfulness, and equity. Through concerted efforts by individuals, communities, and institutions, it is possible to transition from a culture of excess to one characterized by balance and responsibility. This journey towards sustainable and ethical consumption not only promises to alleviate the pressing issues of our time but also to pave the way for a more just and flourishing world for future generations.

Chapter 14: Ethical Bankruptcy - The Erosion of Moral Values

In an era marked by rapid change and increasing complexity, the phenomenon of ethical bankruptcy—where core moral values are disregarded or compromised—presents a formidable challenge to the fabric of society. This decline in ethical standards, observable across various sectors including business, politics, and individual behavior, threatens the foundational trust and cohesion necessary for societal well-being. This chapter seeks to unravel the intricacies of ethical bankruptcy, examining its historical evolution, current manifestations, and the broader implications of this ethical crisis. Through this exploration, we aim to illuminate pathways toward restoring integrity and moral values as guiding principles for contemporary society.

Shifting Moral Landscapes

The journey through humanity's ethical evolution is a testament to the ever-changing nature of societal norms and moral philosophies. This exploration looks into the transformative shifts from the stringent moral frameworks of traditional societies to the nuanced and dynamic ethics characterizing the modern era. By tracing these pivotal historical movements, we gain insight into the forces that have sculpted contemporary notions of morality.

The fabric of moral philosophy has been woven through the ages, influenced by the threads of philosophical insight, religious doctrines, and cultural revolutions. Each era's prevailing conditions—be they economic, social, or technological—have played a critical role in reshaping societal values and ethical standards. This section illuminates the significant developments that have left indelible marks on the landscape of moral thought, offering a foundation to understand the complexities of today's moral values.

One of the most profound influences on ethical evolution has been the advent of major philosophical movements. From the ancient Greeks, who pondered the virtues of the good life, to Enlightenment thinkers, who championed reason and individual rights, philosophy has continually prompted humanity to question and redefine its moral compass. These intellectual traditions have not only spurred debates about the nature of justice, freedom, and equality but have also laid the groundwork for modern ethical frameworks.

Religious teachings have also played an indispensable role in shaping moral landscapes. The moral codes prescribed by world religions have guided billions in their daily lives, influencing societal norms and laws. As civilizations evolved, the interpretation of these religious tenets has adapted, reflecting broader shifts in societal attitudes and behaviors. The Reformation, the Renaissance, and the advent of secularism illustrate how religious influence on moral norms can wax and wane over time.

Cultural milestones and technological advancements have further catalyzed shifts in moral perspectives. The industrial revolution, the abolition of slavery, and the feminist movement are just a few examples of societal shifts that have prompted a reevaluation of ethical principles. In recent times, the digital age has introduced complex moral questions regarding privacy, information ethics, and the role of artificial intelligence in society, highlighting the continuous interplay between technological innovation and ethical considerations.

Today's ethical landscape is marked by its pluralism and adaptability, reflecting the diverse and interconnected world we inhabit. The recognition of human rights, the growing emphasis on environmental stewardship, and the expanding dialogue on social justice issues signify the latest chapters in the ongoing evolution of moral thought. Yet, as global challenges emerge and societal conditions change, the moral landscapes of tomorrow will undoubtedly continue to shift, propelled by the unending quest to define what it means to live a good and just life.

This section not only charts the historical shifts that have shaped our current moral framework but also invites reflection on how today's choices and debates will influence the ethical landscapes of the future. By understanding the roots of our moral values and the forces that have shaped them, we are better equipped to navigate the complex ethical dilemmas of the modern world.

The Impact of Modernity on Ethics

Modernity, with its hallmark traits of technological advancement, rapid urbanization, and expanding globalization, has ushered in profound shifts in ethical norms and the practice of morality. This era of unprecedented change has not only introduced an array of novel ethical dilemmas but has also fostered a pervasive sense of moral relativism, challenging the once solid foundations of right and wrong. The variety of experiences of modern life, marked by the anonymity afforded by dense urban centers and the impersonal interactions characteristic of the digital age, adds layers of complexity to ethical decision-making processes, contributing to a gradual but discernible drift away from traditional moral values.

The relentless pace of technological progress, a defining feature of modernity, has been a double-edged sword. While it has facilitated innovations that improve quality of life and connect the world in ways previously unimaginable, it has also raised unprecedented ethical questions. Issues surrounding data privacy, digital surveillance, and the ethical use of artificial intelligence challenge us to redefine boundaries and establish new moral guidelines in an increasingly digital world. These technological quandaries, coupled with the rapid dissemination of information, amplify the challenges of discerning ethical truths in a sea of relativism.

Urbanization, another cornerstone of modernity, has redefined social structures and interactions. The migration to urban centers has not only physically distanced individuals from traditional community-based support systems but has also introduced a level of anonymity and impersonality into daily life. This detachment from close-knit communities, once the bedrock of ethical guidance and accountability, complicates the navigation of moral landscapes. The dense, bustling environment of cities can dilute a sense of individual responsibility, making ethical considerations seem more abstract and less immediate.

Globalization further complicates the ethical panorama by juxtaposing diverse moral systems and values, leading to an increased questioning of absolute ethical standards. As cultures and economies become more intertwined, the exposure to a multiplicity of ethical perspectives can both enrich understanding and contribute to a sense of moral ambiguity. The global exchange of goods, ideas, and people necessitates a reevaluation of ethical principles in light of cross-cultural interactions and the global impact of local actions.

This confluence of modernity's hallmarks—technological progress, urbanization, and globalization—has not only broadened the scope of ethical dilemmas but has also introduced a degree of moral relativism, challenging individuals to find their ethical footing in a rapidly changing world. The erosion of traditional moral values in the face of modern complexities underscores the need for a renewed focus on ethical education and discourse, aimed at fostering a robust moral compass that can navigate the nuanced realities of the modern era. In this context, the task of defining ethical conduct becomes both more challenging and more critical, as society seeks to balance the benefits of modern advancements with the imperative to maintain a strong ethical foundation.

In Politics

The political landscape frequently serves as a stark tableau of the ethical quandaries that permeate modern society. It is in the realm of politics where the symptoms of ethical malaise often manifest most visibly, through corruption, rampant partisanship at the expense of principled governance, and a troubling erosion of democratic norms. These phenomena highlight a concerning trend where ethical considerations are sidelined in favor of power and expediency. This ethical backsliding within the political sphere not only undermines public trust in governmental institutions but also erodes the very bedrock of effective governance, which ought to prioritize the public good over narrow or short-term political advantages.

Corruption, perhaps the most blatant affront to ethical politics, corrodes the integrity of public institutions by allowing personal or group interests to dictate policy and decision-making. It distorts the democratic process and squanders resources that could be used to address societal needs, effectively stealing from the public to serve the few. The repercussions of corruption extend beyond immediate financial loss, fostering cynicism among citizens about the possibility of clean, fair governance.

Equally detrimental is the rise of extreme partisanship, where loyalty to party supersedes commitment to ethical standards and the collective welfare. This polarized political climate stifles constructive debate, impedes compromise, and hampers the enactment of policies that address complex social issues. When the pursuit of power trumps principled leadership, the fabric of democracy is weakened, and the ability of political systems to function effectively and fairly is compromised.

The erosion of democratic norms—principles such as the rule of law, the sanctity of fair elections, and the importance of civil liberties—represents a profound ethical crisis in politics. As these foundational principles are gradually undermined, often under the guise of achieving politically expedient outcomes, the very essence of democracy is threatened. This erosion risks entrenching authoritarian tendencies and diminishing the capacity for democratic renewal.

These trends within the political sphere reflect a broader ethical challenge that transcends individual misconduct to encompass systemic issues that require collective attention and action. Rebuilding public trust and restoring the integrity of political institutions demand a recommitment to ethical governance. This entails not only holding individuals accountable for unethical behavior but also fostering a political culture that values transparency, accountability, and a steadfast dedication to the public interest. Addressing the ethical challenges in politics is crucial for ensuring that governance truly serves its foundational purpose: to facilitate the well-being of society and safeguard the democratic principles that underpin it.

In Business Practices

The realm of business has not been immune to ethical lapses, witnessing a spectrum of misconduct that spans financial improprieties, labor exploitation, and environmental negligence. These unethical practices, often motivated by an unrelenting pursuit of profit, carry immediate detrimental impacts on individuals, communities, and ecosystems. Moreover, they prompt a critical examination of the long-term viability and moral integrity of such business models. The fallout from these ethical failures in the business sector underscores a pressing need for a paradigm shift towards more responsible and ethical corporate conduct.

Financial misconduct, encompassing a range of malpractices from fraud to insider trading, erodes trust in the financial markets and undermines the principle of fair competition. These actions not only inflict direct economic harm but also compromise the overall stability of the financial system. The repercussions are far-reaching, affecting not just the perpetrators and their immediate victims but also shaking investor confidence and public trust in financial institutions.

Exploitation of labor is another stark example of ethical compromise in the business world. Practices such as unfair wages, unsafe working conditions, and child labor reflect a blatant disregard for human rights and dignity. These practices are not only morally reprehensible but also counterproductive, as they contribute to social instability and perpetuate cycles of poverty and inequality. The global nature of supply chains further complicates these issues, often obscuring accountability and making it challenging to enforce ethical standards across borders.

Environmental negligence—manifested through practices such as unsustainable resource extraction, pollution, and contribution to climate change—highlights a critical ethical oversight in prioritizing short-term gains over the planet's long-term health. The environmental degradation resulting from such practices poses existential threats and calls into question the responsibility of businesses to the broader ecosystem and future generations.

The cumulative social and environmental costs associated with unethical business practices illuminate the imperative for a shift towards more sustainable and principled corporate behavior. This transition requires a holistic reevaluation of corporate values and objectives, where profit-making is balanced with ethical responsibilities to society and the environment. Implementing ethical business practices involves embracing transparency, accountability, and a genuine commitment to doing no harm as foundational principles.

Moving forward, the business community must recognize that true success and sustainability lie not in maximizing short-term profits but in fostering long-term well-being for people and the planet. By integrating ethical considerations into decision-making processes, businesses can lead the way in building a more just and sustainable world. This approach not only mitigates the risks of ethical failures but also positions companies to thrive in an increasingly conscientious market, where consumers and investors alike demand responsible corporate conduct.

The Role of Individual Choices

The gradual erosion of ethical standards finds a potent expression in the realm of personal conduct, where the daily choices of individuals are both a reflection and a constituent of the broader ethical landscape. Navigating the pressures of a highly competitive, consumer-driven society often places individuals at a crossroads, where the temptation to compromise on integrity and honesty can clash with deeply held values and principles. This tension underscores the significant role that individual choices play in either perpetuating or challenging the ethical status quo.

The rapid pace of life and the relentless pursuit of success in contemporary society can blur ethical boundaries, making it increasingly difficult for individuals to uphold their moral commitments. Decisions that might once have seemed black and white become muddied by the complexities of modern life, where the consequences of actions are not always immediately apparent, and the stakes are perceived to be higher than ever. In such an environment, the justification for ethical compromises can be readily found in the perceived necessity of keeping pace with societal expectations or achieving personal goals.

Moreover, the digital age introduces additional layers of complexity to ethical decision-making. The anonymity afforded by online interactions and the omnipresence of information can dilute the sense of personal accountability and obscure the impact of one's actions on others. Social media platforms, where personal and public spheres often intersect, can further complicate the ethical landscape, presenting scenarios where the line between right and wrong is not easily discerned.

In these intricate contexts, making ethical choices requires not only a clear understanding of one's values but also the courage to act in accordance with them, even when faced with potential personal or professional costs. It demands a heightened awareness of the broader consequences of our actions and a commitment to principles over convenience.

The role of individual choices in shaping the ethical fabric of society cannot be overstated. Each decision contributes to a collective ethical climate, influencing norms and setting precedents for others. As such, fostering an environment that encourages ethical vigilance and integrity starts with individuals choosing to act responsibly and conscientiously, despite the pressures and temptations of modern life.

Recognizing the power of personal choices in the ethical domain underscores the importance of cultivating moral clarity and resilience. By prioritizing ethical considerations in our daily decisions, individuals can play a crucial role in reversing the trend of ethical erosion and contribute to a more principled and just society. This commitment to ethical living, while challenging, is essential for ensuring that the values we espouse are mirrored in the world we help to create.

The Impact on Social Relationships

The gradual decline in ethical standards carries profound implications for the fabric of social relationships, extending well beyond the realm of individual conduct. At the heart of these relationships lies trust, an indispensable component of social cohesion and collective well-being. As ethical erosion deepens, trust is severely compromised, giving way to a pervasive atmosphere of cynicism and skepticism. People begin to question the integrity and motives of those around them, whether in personal interactions, business transactions, or political discourse. This decline in trust fundamentally alters the nature of social connections, impacting everything from the most intimate interpersonal relationships to the broader constructs of societal engagement.

The repercussions of diminished trust are far-reaching, affecting the very foundation upon which communities and societies are built. As individuals grow more guarded, their willingness to participate in communal activities, share resources, and support collective endeavors wanes. This retreat into skepticism and self-preservation undermines the social contract—the implicit agreement among members of a society to cooperate for social benefits. When trust erodes, the social contract frays, leading to a weakened sense of community and a decline in the quality of social interactions.

This atmosphere of distrust can have a cascading effect, exacerbating social divides and fueling isolation. Without trust, the fabric that binds individuals into a cohesive community becomes threadbare, leaving people feeling disconnected and alone. Such conditions foster an environment where misinformation can thrive and divisive rhetoric can sow discord, further impeding the ability to build or maintain meaningful relationships.

The impact on social relationships extends to the realm of civic engagement and democratic participation. When citizens distrust their leaders, institutions, and even each other, civic life suffers. People are less likely to vote, engage in constructive dialogue, or partake in community service, which are critical components of a vibrant and functioning democracy. This disengagement signifies not just a loss of faith in the system but also a missed opportunity for collective problem-solving and social progress.

To counteract the corrosive effects of ethical erosion on social relationships, efforts must be made to rebuild trust at all levels of society. This endeavor requires transparency, accountability, and a renewed commitment to ethical behavior from individuals, institutions, and leaders alike. By fostering environments where honesty and integrity are valued and rewarded, it is possible to rekindle trust and strengthen the social bonds that are essential for a healthy and cohesive society.

The restoration of trust is not a quick or easy process, but it is a necessary one for the repair and revitalization of social relationships. Through deliberate actions and a collective commitment to ethical principles, society can navigate the challenges of ethical erosion and emerge with a stronger, more resilient foundation of trust. This, in turn, can pave the way for a more connected, engaged, and cohesive community, where social relationships thrive on mutual respect and understanding.

Education and Ethical Literacy

Addressing the challenge of ethical erosion and the revitalization of moral standards within society hinges significantly on education. The integration of ethical education into curricula from an early age represents a pivotal strategy in shaping a generation equipped with the competencies necessary to navigate the intricate moral dilemmas of contemporary life. Ethical literacy, defined by the ability to critically analyze complex situations and make informed moral judgments, stands as a crucial pillar in fostering a societal culture deeply rooted in values of integrity, responsibility, and ethical conduct.

The importance of ethical literacy extends beyond academic achievement; it is about preparing individuals to engage with the world in a thoughtful, principled manner. This involves instilling a nuanced understanding of ethical concepts, including the diversity of moral perspectives, the reasoning behind ethical decisions, and the impact of actions on individuals and communities. By embedding ethical considerations into the fabric of education, students learn to appreciate the significance of moral values not only in personal decision-making but also in the broader contexts of social justice, environmental stewardship, and global citizenship.

Ethical education encourages the development of critical thinking skills, empathy, and a sense of accountability—qualities that are indispensable in confronting the ethical challenges of modern society. It provides a framework for students to question and reflect upon the ethical dimensions of various issues, from everyday interpersonal interactions to complex global dilemmas. Through discussion, debate, and practical application, ethical education cultivates an environment where students are encouraged to form their moral viewpoints, recognize ethical conflicts, and consider the consequences of their decisions.

Incorporating ethical literacy into the educational system also involves equipping educators with the tools and resources to effectively teach these concepts. This includes training in ethical theory, case studies that highlight real-world ethical dilemmas, and pedagogical strategies that foster active engagement and critical reflection. Moreover, creating a school culture that embodies ethical principles—in policies, interactions, and governance—reinforces the lessons taught in the classroom, showing students that ethics are not just theoretical concepts but practical guides for conduct.

Institutional Reforms and Corporate Responsibility

The battle against ethical erosion and the quest for a restoration of moral values significantly hinge on the actions of institutions—both in the political realm and the corporate world. These entities wield immense influence over societal norms and possess the capacity to either uphold or undermine ethical standards through their practices. Institutional reforms that emphasize transparency, accountability, and sustainability are crucial in modeling ethical behavior and establishing benchmarks that others might follow. By fostering a culture of corporate responsibility that extends beyond mere profit-making to encompass societal welfare and environmental stewardship, institutions can lead by example, demonstrating the compatibility of ethical conduct with business success.

Political institutions, tasked with governance and the public trust, must champion reforms that enhance transparency and accountability. This involves implementing stringent anti-corruption measures, ensuring open and fair political processes, and fostering public engagement in democratic practices. Such reforms are fundamental in rebuilding trust and integrity within the political sphere, encouraging citizens to participate more actively in governance and civic life.

In the corporate sector, the concept of corporate social responsibility (CSR) serves as a beacon for ethical business practices. CSR goes beyond traditional business goals to include a commitment to positive societal impact and environmental sustainability. This involves reevaluating corporate strategies, supply chains, and operations to ensure they align with ethical principles such as fairness, respect for human rights, and ecological conservation. By prioritizing these values, corporations can contribute to societal well-being while still achieving profitability.

Encouraging corporate responsibility entails recognizing the broader impact of business operations on society and the environment. This recognition should spur efforts to minimize negative impacts while maximizing positive contributions to communities and ecosystems. For example, adopting green technologies, supporting fair labor practices, and engaging in philanthropic activities are tangible ways companies can exhibit corporate responsibility.

Institutional reforms and corporate responsibility are not merely internal concerns but also involve engagement with stakeholders—customers, employees, communities, and governments. Open dialogue and collaboration with these groups can enhance ethical practices, ensuring that business operations reflect the values and needs of a broad spectrum of society. Such engagement also provides a platform for accountability, where institutions can be held to their ethical commitments and encouraged to make continuous improvements.

Ultimately, the path towards more ethical business practices and governance requires a paradigm shift in how institutions view their roles in society. By integrating ethical considerations into decision-making processes and prioritizing long-term societal welfare over short-term gains, political and corporate institutions can contribute significantly to a more ethical, just, and sustainable world. This shift not only benefits society at large but also strengthens the institutions themselves, fostering resilience, enhancing reputation, and building trust among stakeholders.

As society grapples with the complexities of ethical erosion, the role of education in promoting ethical literacy becomes ever more critical. By prioritizing ethical education, we invest in the development of morally conscious individuals who are capable of contributing to a more just, compassionate, and ethical world. This approach not only addresses the immediate need for a renewed emphasis on ethical standards but also lays the groundwork for enduring change, ensuring that future generations possess the moral clarity and courage to face the ethical challenges of their times.

Chapter 15 Educational Erosion - Failing Systems and Lost Potentials

The concept of educational erosion captures the gradual decline in the quality and equity of education systems worldwide. Characterized by underfunding, outdated teaching methods, and widening disparities, this erosion poses significant threats to individual development and societal progress. As education forms the cornerstone of personal empowerment and economic advancement, understanding and addressing the factors contributing to its decline is crucial. This chapter aims to dissect the complexities of educational erosion, from its root causes to its varied manifestations, and explore pathways towards revitalizing education systems to fulfill their potential as engines of growth and equality.

Underfunding and Resource Gaps

A primary driver of educational decline is the chronic underfunding of schools, especially in low-income areas. This pervasive issue stems from a complex web of socio-economic factors, policy decisions, and sometimes, neglect. At its core, underfunding manifests as a significant gap in resources available to schools, directly impacting their ability to deliver quality education.

The consequences of underfunding are far-reaching and multilayered. Firstly, it leads to teacher shortages. Qualified educators are the backbone of any education system, playing a crucial role in shaping students' learning experiences and outcomes. However, underfunded schools often struggle to attract and retain skilled teachers due to uncompetitive salaries, poor working conditions, and limited professional development opportunities. This not only degrades the quality of education but also places an undue burden on the teachers who remain, often leading to burnout and further exacerbating the shortage.

Inadequate learning materials and resources are another critical issue. Textbooks, laboratory equipment, technology, and other educational materials are essential for a comprehensive learning experience. Unfortunately, schools suffering from financial neglect may not have sufficient or up-to-date materials, forcing teachers to work with what they have, which is often insufficient to meet curriculum standards or engage students effectively.

The physical state of educational facilities in underfunded districts is a glaring testament to the neglect. Crumbling infrastructure is not just an aesthetic issue; it poses safety risks and creates an environment that can be demoralizing for students and staff alike. Leaking roofs, malfunctioning heating and cooling systems, and outdated facilities do not foster a conducive learning environment.

The disparity in funding between schools in affluent areas and those in impoverished districts is perhaps the most visible indicator of inequality within the education system. This gap not only highlights the financial aspect but also reflects broader societal disparities that impact education. Wealthier districts can often supplement public funding with additional resources from local taxes or fundraising efforts, further widening the gap.

This systemic underfunding denies countless students the opportunity for a quality education, setting up a cycle of disadvantage that can affect their future employment, income, and overall life opportunities. Addressing this issue requires a dynamic approach, including policy reform, increased investment in education, and a commitment to equity that ensures all students, regardless of their socio-economic background, have access to the tools and resources they need to succeed.

Outdated Curricula and Pedagogies

One of the most significant hurdles in contemporary education is the persistence of outdated curricula and teaching methods that are misaligned with the requirements of the modern world. In numerous education systems, there's a noticeable reliance on rote learning and an overemphasis on standardized testing. This approach not only stifles creativity but also sidelines the development of critical thinking skills and practical knowledge that are indispensable in today's digitally driven and rapidly evolving society.

The emphasis on memorization over understanding reflects a dated view of education, one where students are seen more as vessels to be filled rather than individuals to be developed. This methodology neglects the nurturing of curiosity, problem-solving skills, and the ability to apply knowledge in real-world contexts. As a result, while students may excel in replicating information, they often struggle with synthesizing and applying this knowledge creatively and effectively.

The curricula in many schools have not kept pace with the technological advancements and societal changes that define the 21st century. Subjects that are crucial for navigating today's complex, globalized world, such as digital literacy, environmental science, and cross-cultural communication, are often absent or only marginally covered. This gap leaves students underprepared for the realities of the modern workforce, where such competencies are increasingly valued and required.

The use of outdated pedagogies compounds the problem. Traditional lecture-based teaching methods dominate, with limited opportunities for interactive, experiential learning. This one-size-fits-all approach fails to account for diverse learning styles and the individual needs of students, making it difficult for some to engage with and absorb the material fully. The lack of personalization and flexibility in teaching methods can hinder the educational progress of students who might thrive under different instructional strategies.

This disconnect between the content and methods of education and the demands of contemporary life creates a significant challenge. It leaves students ill-equipped to tackle the complexities of the workforce and civic engagement, impacting not only their personal and professional success but also the societal progress at large. Bridging this gap requires a comprehensive overhaul of curricula and teaching methodologies. Education systems need to embrace a more holistic approach that prioritizes critical thinking, creativity, and adaptability, preparing students not just to succeed in the present but to innovate for the future.

Socioeconomic Disparities

The interplay between socioeconomic status and educational outcomes is both profound and complex, casting long shadows over the landscape of educational equity. Students hailing from low-income families frequently confront a barrage of barriers that can significantly hamper their educational journey. Among these impediments, limited access to quality early childhood education stands out as a foundational disadvantage. Early childhood education is critical for cognitive and social development, setting the stage for future learning. However, due to financial constraints, children from less affluent backgrounds often miss out on these enriching experiences, starting their educational journey at a disadvantage.

Under-resourced schools disproportionately located in low-income areas further exacerbate these disparities. Such schools often lack adequate funding, leading to overcrowded classrooms, insufficient learning materials, and fewer qualified teachers. The physical condition of these schools can also be subpar, all of which contribute to a learning environment that is less conducive to student success. The disparity in educational resources and opportunities means that students from wealthier backgrounds are afforded a head start, while their less affluent peers are left playing catch-up.

The socioeconomic challenges extend beyond the classroom. Many students from low-income families are compelled to take on work or household responsibilities from a young age. These obligations can significantly detract from their time and energy for academic pursuits, homework, and extracurricular activities that are vital for a well-rounded education. Such students often have to prioritize immediate financial contributions to their family over long-term educational achievements, further entrenching the cycle of poverty.

These disparities do not just limit the individual potential of students but also perpetuate broader cycles of poverty and inequality. Education is widely recognized as a powerful vehicle for social mobility, offering a pathway out of poverty. However, when access to quality education is skewed by socioeconomic status, it undermines this potential for upliftment, entrenching existing social divides.

Addressing these socioeconomic disparities requires a concerted effort from policymakers, educators, and communities. Initiatives aimed at providing equitable access to quality early childhood education, improving funding and resources for under-resourced schools, and supporting students with extracurricular and academic opportunities can help level the playing field. Ultimately, bridging these gaps is not just about improving individual outcomes but about fostering a more equitable and cohesive society.

Barriers to Education for Marginalized Groups

Education, ideally a universal right, remains inaccessible for many within marginalized communities due to a constellation of barriers rooted in discrimination, cultural biases, and systemic inequities. These barriers, deeply embedded in the fabric of societies, significantly hinder the educational attainment and development of racial and ethnic minorities, girls and women, and students with disabilities, among others. The challenges these groups face are not only diverse but also exacerbate the existing inequalities in education and beyond.

For racial and ethnic minorities, discrimination can manifest in various aspects of the educational experience, from unequal access to quality schools to biased disciplinary practices and curricular content that does not reflect their cultural heritage. Such systemic bias not only impacts academic outcomes but also affects students' self-esteem and sense of belonging within educational institutions.

Girls and women, particularly in regions where gender norms restrict their rights and opportunities, often encounter significant obstacles to education. These can range from explicit prohibitions on female education to more subtle forms of discrimination, such as a lack of sanitary facilities that accommodate their needs, or cultural expectations prioritizing their roles in domestic work over academic pursuits. These barriers significantly limit their educational opportunities and future prospects.

Students with disabilities face challenges related to accessibility, both in physical infrastructure and in teaching methodologies and materials. The lack of inclusive educational practices and resources can prevent these students from fully participating in and benefiting from education. Furthermore, societal stigma and misunderstanding about disabilities further complicate their educational journey.

Addressing the barriers faced by marginalized communities demands a commitment to targeted interventions and inclusive policies. This entails not only removing the structural obstacles to education but also fostering an educational environment that celebrates diversity, promotes inclusion, and provides all students with the support they need to succeed. Implementing inclusive education policies, ensuring equitable funding and resources, and adopting culturally responsive teaching practices are critical steps toward dismantling the barriers and ensuring equal educational opportunities for every student, regardless of their background.

Comparative International Perspectives

Educational erosion, an issue, presents itself distinctly across the globe, shaped by unique combinations of economic conditions, cultural contexts, and policy frameworks. From the developed corridors of the Western world to the developing landscapes of the Global South, each region faces its own set of challenges and successes in education. This section embarks on a comparative analysis, shedding light on how various factors contribute to educational disparities and what can be learned from the diverse educational models employed around the world.

In many developing countries, the crux of educational erosion lies in chronic underfunding and inadequate infrastructure. Schools in rural Africa, for instance, often lack basic facilities such as classrooms, textbooks, and even sanitation facilities, which significantly impede the learning process. Moreover, teacher shortages and a lack of professional training further exacerbate the problem, leaving many children without the quality education they deserve.

Conversely, developed nations, while generally well-equipped in terms of infrastructure and resources, face their own set of challenges. In the United States and parts of Europe, issues such as socioeconomic disparities, racial and ethnic segregation, and uneven distribution of resources lead to significant gaps in educational outcomes. The increasing focus on standardized testing and curricula that prioritize rote learning over critical thinking skills has sparked debates about the quality and relevance of education in the modern world.

Countries in Asia present another variation of educational challenges and achievements. Nations like South Korea and Japan have achieved remarkable success in standardized testing and higher education enrollment rates. However, this success often comes at the cost of intense academic pressure, mental health issues, and a lack of emphasis on creativity and social skills in the curriculum. Meanwhile, in many parts of South Asia, rapid population growth and gender disparities in education access remain significant hurdles.

Looking at the Nordic countries, we see a different educational model characterized by strong government support, inclusive policies, and a focus on holistic development. These nations prioritize equitable access to education, teacher autonomy, and student well-being, resulting in high levels of literacy and educational attainment. The Nordic model underscores the importance of viewing education as a collective societal investment rather than a commodity.

From these comparative international perspectives, several lessons emerge. First, the significance of government investment and policy frameworks in shaping educational outcomes cannot be overstated. Second, addressing educational erosion requires a comprehensive approach that considers economic, social, and cultural factors. Finally, while there is no one-size-fits-all solution, successful educational models often share common features such as inclusivity, a focus on teacher training, and curricula that foster critical thinking and creativity.

This analysis reveals that educational erosion is a complex issue with diverse manifestations around the world. By drawing lessons from various educational models, countries can develop strategies tailored to their specific contexts, aiming for an educational system that not only imparts knowledge but also prepares students for the challenges of the modern world.

The Role of Globalization and Technology

The dual forces of globalization and technology have a profound impact on education, presenting a complex array of challenges and opportunities. As the global economy evolves, it demands a workforce equipped with a new set of skills and competencies, including digital literacy, critical thinking, and the ability to work across cultural boundaries. This shift places significant pressure on educational systems worldwide to adapt their curricula and teaching methods to prepare students for the realities of a interconnected world.

Globalization has intensified the need for education systems to foster global citizenship, understanding, and cooperation. Students today must be prepared to navigate an increasingly diverse and interconnected global landscape. This necessitates a shift in educational priorities towards a more holistic approach that includes not only traditional academic skills but also emotional intelligence, cultural awareness, and the capacity for innovation and adaptability.

On the other side of the coin, technology offers transformative potential for education. Digital learning tools, online educational resources, and the global connectivity afforded by the internet can greatly enhance the delivery and accessibility of education. Technological advancements have made it possible to personalize learning, cater to diverse learning styles, and provide access to world-class resources and experts regardless of geographical location. This democratization of access to knowledge is a significant step forward in overcoming traditional barriers to education, such as geographical isolation, resource constraints, and socio-economic disparities.

However, the integration of technology in education is not without its challenges. The digital divide remains a significant barrier, with access to technology and the internet still highly unequal both within and between countries. Moreover, there is the risk that an overreliance on technology could exacerbate existing inequalities and neglect the development of interpersonal skills and critical thinking.

The role of globalization and technology in education is thus a double-edged sword. While they bring about challenges that require educational systems to adapt, they also offer unprecedented opportunities for enhancing education and making it more accessible and relevant to the needs of the global economy. To fully leverage these opportunities, it is crucial for policymakers, educators, and stakeholders to work towards equitable access to technology, adapt curricula to the demands of the 21st century, and ensure that education remains a holistic process that prepares students not just for the workforce, but for life in a globalized world.

Innovations in Funding and Resource Allocation

To effectively combat educational erosion, a radical rethinking of funding and resource allocation strategies is imperative. The traditional models, often constrained by bureaucratic limitations and a one-size-fits-all approach, have proven insufficient in addressing the complex needs of today's diverse student populations. Innovative financing solutions and equitable resource distribution mechanisms are essential to revitalizing educational systems and ensuring that all students have access to quality education.

Public-private partnerships (PPPs) stand out as a promising strategy for enhancing educational funding. By leveraging the strengths and resources of both sectors, PPPs can provide additional funding streams, introduce innovative teaching methodologies, and foster the integration of technology in classrooms. These partnerships not only bring in financial resources but also encourage the sharing of best practices and expertise, driving improvements in educational quality and efficiency.

Community involvement in schools is another critical element in the innovation equation. By engaging local businesses, non-profits, and families in the education process, schools can access a wider range of resources and support. This may include mentorship programs, donations of materials and technology, or volunteer time. Community engagement not only augments resources but also strengthens the ties between schools and their communities, creating a supportive ecosystem that values and prioritizes education.

Targeted investments in high-need areas represent a focused approach to addressing educational disparities. By directing funds towards schools in underprivileged or underserved communities, it's possible to make significant strides in leveling the playing field. These investments can support infrastructure improvements, enhance teacher training, and provide students with the learning materials and technology they need to succeed.

The allocation of resources based on student needs rather than traditional metrics, such as standardized test scores or attendance rates, offers a more equitable approach to funding. This method acknowledges the unique challenges faced by different student populations and seeks to allocate resources in a manner that addresses those specific needs. For instance, schools serving a large number of students from low-income families or students with disabilities may require additional support to ensure that these students have equal opportunities to succeed.

Adopting innovative approaches to funding and resource allocation is not without challenges. It requires a shift in mindset from all stakeholders, a willingness to experiment with new models, and a commitment to equity and inclusivity. However, the potential rewards—more equitable education systems that serve the needs of all students and prepare them for the challenges of the future—are well worth the effort. By embracing innovation, policymakers and educators can ensure that education remains a dynamic force for empowerment and societal progress.

Reforming Curricula and Pedagogies

Addressing the current educational challenges and preparing students for the complexities of the 21st century demand a fundamental overhaul of both curricula and pedagogies. The objective is to move beyond traditional educational models focused on memorization and standardized testing, towards a more dynamic, inclusive, and forward-thinking approach. This reform is crucial for developing the skills and competencies that students need to thrive in an ever-changing global landscape.

Integrating Technology and Digital Literacy: In today's digital age, proficiency in technology and digital literacy is not optional but essential. Educational curricula must be reformed to incorporate these elements comprehensively, ensuring students are comfortable and competent with digital tools and platforms. This integration extends beyond just using technology in the classroom; it involves teaching students to critically assess digital information, understand the ethical implications of digital technologies, and leverage these tools creatively and effectively in various aspects of their lives.

Emphasizing Critical Thinking and Problem-solving Skills: The ability to think critically, analyze complex information, and devise innovative solutions to problems is more valuable than ever. Curricula should be designed to cultivate these skills, encouraging students to question, critique, and engage deeply with the material they learn. This shift requires moving away from rote learning towards more inquiry-based, project-based, and experiential learning methods that challenge students to apply their knowledge in real-world contexts.

Adopting Flexible and Student-centered Teaching Methods: Teaching methodologies need to be as diverse as the students they aim to educate. Adopting a more flexible and student-centered approach allows educators to tailor their teaching to meet the varied needs, interests, and learning styles of their students. Such methods foster a more engaging and supportive learning environment, empowering students to take an active role in their education and develop a love for learning.

Encouraging Lifelong Learning: Recognizing that education does not end with formal schooling is crucial in today's rapidly evolving world. Curricula and pedagogies should instill the value of lifelong learning, preparing students to continuously acquire new knowledge and skills throughout their lives. This perspective not only benefits individuals but also contributes to the overall adaptability and resilience of societies facing constant technological and societal changes.

The challenge of educational erosion is indeed complex, touching on various aspects of society and individual development. Overcoming it requires a concerted effort that addresses the issue from multiple angles. By investing in innovative funding and resource allocation, reforming curricula and teaching methods, and fostering an inclusive and adaptive educational environment, we can revitalize education systems globally. Such comprehensive efforts are vital for ensuring that education continues to serve as a pivotal mechanism for personal empowerment, social equity, and the advancement of communities and nations alike.

Chapter 16: The Illusion of Progress - Questioning Technological Salvation

In our era defined by rapid technological advancements, the belief in technology as a savior for humanity's myriad challenges has become pervasive. This conviction, often termed "technological salvation," posits that through innovation and scientific progress, society can address its most pressing issues, from climate change to health crises. However, this belief warrants a critical examination of the assumptions it rests upon and the potential it has to overlook the complex ethical, social, and environmental implications of unchecked technological development. By delving into the nuances of technological progress, this chapter aims to unveil the illusion of progress and advocate for a more conscientious approach to harnessing technology for the genuine betterment of society.

Historical Context of Techno-optimism

Techno-optimism, the belief in technology as a catalyst for positive societal change, is not a recent phenomenon. Its seeds were sown centuries ago, deeply rooted in the milestones of human achievement that marked each era of progress. The Industrial Revolution, spanning the 18th and 19th centuries, is often cited as the initial spark for this faith in technological progress. It was a period characterized by significant advancements in manufacturing and production processes, which not only revolutionized industries but also profoundly transformed societies. These changes brought about unprecedented economic growth and improvements in living standards for many, cementing the idea that technological progress was inherently beneficial.

As the world moved into the 20th century, this belief was further bolstered by a series of technological feats that seemed to confirm the unlimited potential of human innovation. The invention of antibiotics, the harnessing of nuclear energy, and the advent of computing and the Internet are just a few examples of how technology continued to be seen as a powerful force for good. Each breakthrough promised to solve pressing societal issues, from health crises to enhancing global communication and information access.

This historical narrative, however, tends to overlook the more complex and often darker side of technological progress. While technology has indeed brought about significant benefits, it has also catalyzed societal upheaval and widened inequalities. The same Industrial Revolution that improved lives also ushered in harsh working conditions, urban overcrowding, and environmental degradation. Similarly, the 20th-century technological advancements, while transformative, introduced new ethical dilemmas—nuclear weapons brought about new existential threats, and digital technologies raised concerns over privacy and data security.

The environmental impact of relentless technological advancement has become increasingly apparent. The extraction of natural resources, pollution, and the creation of non-biodegradable waste are just some examples of how technology can contribute to environmental degradation. These issues highlight the complexity of technological progress, challenging the simplistic narrative of unequivocal societal improvement.

In reflecting on the historical context of techno-optimism, it becomes clear that while technology holds immense potential for good, its impacts are varied and often contradictory. The narrative of technological advancement as synonymous with societal progress needs to be critically examined, recognizing both its achievements and its capacity to exacerbate or create new challenges. This nuanced understanding is essential for navigating the future of technological development in a way that genuinely benefits society as a whole, without overlooking the ethical, social, and environmental considerations that accompany innovation.

Challenging the Notion of Infallible Progress

The narrative of technological progress as a linear path leading humanity towards an ever-brighter future is a pervasive one. It's a belief deeply ingrained in the collective consciousness, fueled by remarkable innovations that have undeniably improved many aspects of human life. However, this narrative of infallible progress, where each technological advancement is a step closer to utopia, demands scrutiny. History presents us with a more nuanced story, one that includes chapters filled with cautionary tales of how technology, despite its potential for good, can have far-reaching negative impacts.

The environmental consequences of industrialization serve as a stark example of the dual-edged nature of technological progress. While it has propelled economic growth and enhanced our ability to manipulate the material world, it has also led to deforestation, pollution, and climate change. These environmental issues are not mere side effects but significant threats that jeopardize the very fabric of life on Earth. The promise of industrial and technological advancement has, in many cases, come at the expense of the planet's health and biodiversity.

In the realm of digital technology, the advancements have been nothing short of revolutionary. Yet, these technologies have ushered in a new era of privacy concerns and surveillance capabilities that were once the stuff of science fiction. The ability to collect, store, and analyze vast amounts of personal data has transformed not only how we interact with the world but also how we are monitored. The proliferation of digital surveillance technologies has raised significant ethical questions about the right to privacy and the extent to which individuals are unwittingly subjected to observation and data extraction.

The digital divide—the gap between those who have access to modern information and communication technology and those who do not—exemplifies how technological progress can exacerbate socio-economic inequalities. Rather than leveling the playing field, the rapid pace of digital innovation has often widened the gap between the haves and the have-nots, both within societies and globally. Access to technology has become a critical determinant of economic opportunity, education, and social mobility, leaving those without access further behind.

These examples underscore the necessity of adopting a critical stance towards the notion of infallible technological progress. While technology holds the potential to address some of humanity's most pressing challenges, its impacts are complex and layered. The unconditional optimism that has long characterized our approach to technological innovation must be tempered with a conscientious examination of its potential downsides. It is imperative to question, scrutinize, and debate the trajectory of technological development, ensuring that it serves not just the interests of a few but the well-being of society at large. In doing so, we can navigate the path of progress with wisdom, ensuring that the advancements we embrace today do not become the dilemmas of tomorrow.

Surveillance and Privacy

In the digital age, surveillance has evolved from a straightforward observation to a complex web of data collection, analysis, and monitoring that spans the globe. The proliferation of digital technologies has significantly enhanced the capabilities of governments, corporations, and even individuals to conduct surveillance at an unprecedented scale. While these advancements are often justified in the name of security, they encroach upon the sanctity of personal privacy, igniting a debate that lies at the heart of contemporary ethical dilemmas.

Surveillance cameras, once confined to high-security areas, now dot the landscapes of cities and neighborhoods worldwide. Their silent gaze captures the movements of millions, often without explicit consent or awareness. This omnipresent surveillance apparatus is augmented by sophisticated data mining practices that sift through vast amounts of personal information gathered from online activities. Every click, search, and interaction is subject to scrutiny, painting detailed portraits of individuals' lives, preferences, and behaviors.

The rise of digital monitoring extends beyond visual surveillance and data mining. Social media platforms, smartphones, and various Internet of Things (IoT) devices contribute to a digital footprint that is vast and continuously expanding. These technologies, while offering unprecedented connectivity and convenience, also serve as conduits for surveillance, channeling personal data into databases that are ripe for analysis and, potentially, exploitation.

This extensive surveillance network raises pressing ethical questions about the balance between security and privacy. At what point does the quest for safety infringe upon the fundamental right to privacy? The ethical considerations surrounding surveillance are not merely academic; they strike at the core of personal freedom and autonomy. The ability to move, communicate, and express oneself without the specter of constant surveillance is a cornerstone of a free society. Yet, as surveillance technologies become more embedded in everyday life, the boundaries of acceptable monitoring become increasingly blurred.

The ethical examination of surveillance in the digital age must grapple with these complex issues, seeking to define the limits of technological use in monitoring individuals. It requires a delicate balance, weighing the legitimate needs of security against the inviolable rights of individuals to privacy and autonomy. As digital technologies continue to advance, so too must our understanding of their implications for privacy and the ethical frameworks that govern their use. The challenge lies in crafting policies and practices that safeguard personal freedoms without compromising the benefits that these technologies bring to society. This ongoing dialogue is crucial for ensuring that the digital future remains one where privacy is preserved as a fundamental human right, even in the face of ever-expanding surveillance capabilities.

Artificial Intelligence and Moral Responsibility

The integration of artificial intelligence (AI) and automation into the fabric of society is advancing at an unprecedented pace, heralding a new era of innovation and efficiency. However, this technological revolution brings with it profound ethical dilemmas that demand careful consideration. One of the most pressing concerns is the impact of AI and automation on employment. The specter of job displacement looms large, with automation threatening to render obsolete entire categories of jobs that have been staples of the workforce for generations. This shift not only poses a risk to economic stability but also raises fundamental questions about identity, purpose, and the value of human labor in a post-automated world.

Beyond the economic implications, the delegation of decision-making to AI systems introduces complex ethical challenges. As algorithms begin to make choices that affect people's lives, from judicial sentencing recommendations to medical diagnoses, the fairness and integrity of these decisions come under scrutiny. The opacity of many AI systems—often referred to as "black boxes" due to their inscrutable decision-making processes—exacerbates these concerns. Without transparency, understanding how decisions are made, and on what basis, becomes nearly impossible, eroding trust and accountability.

The issue of accountability for AI-driven actions is another critical area of ethical debate. When an AI system makes a decision that has negative consequences, determining who—or what—is responsible is not straightforward. The distributed nature of AI development, from the data used to train algorithms to the designers and programmers who build them, complicates the assignment of blame or liability. This ambiguity challenges our traditional concepts of moral and legal responsibility, necessitating new frameworks and regulations that can address the unique characteristics of AI.

To navigate these ethical dilemmas, there is a pressing need for robust ethical frameworks that guide the development and deployment of AI and automation technologies. Such frameworks should prioritize the protection of human rights, ensure fairness and transparency in AI decision-making, and establish clear guidelines for accountability. Moreover, engaging a diverse range of stakeholders in the creation of these frameworks is crucial to address the wide reaching impacts of AI across different segments of society.

The development of AI and automation presents an opportunity to reimagine the future, promising immense benefits if guided by a strong ethical compass. As we stand on the brink of this new technological frontier, it is incumbent upon policymakers, technologists, and society at large to ensure that these innovations are harnessed responsibly, with a keen eye towards the moral implications of entrusting machines with decisions that shape human lives.

Resource Depletion and Pollution

In an era marked by unprecedented technological advancement, the environmental impact of our digital and electronic consumption is a growing concern. The lifecycle of technological products—from production and consumption to disposal—casts a long shadow on the environment, highlighting the urgent need for sustainable practices. At the core of this issue are the extraction and depletion of rare earth metals, critical components in everything from smartphones to electric vehicles. The mining of these materials is not only energy-intensive but also leads to significant environmental degradation, including habitat destruction, water pollution, and greenhouse gas emissions.

The environmental costs extend beyond the extraction of raw materials. The energy consumption of the world's data centers, which power the internet, cloud computing, and a multitude of digital services, is staggering. These facilities require vast amounts of electricity, much of which is still sourced from non-renewable energy, contributing to carbon emissions and exacerbating climate change. As the digital economy continues to grow, finding ways to power these essential infrastructures sustainably becomes increasingly critical.

Electronic waste (e-waste) represents another significant environmental challenge. The rapid pace of technological innovation and the consequent obsolescence of electronic devices have led to a surge in e-waste. This not only includes consumer electronics like smartphones and laptops but also a wide range of other technologies that end their lifecycle in landfills. Improper disposal of e-waste not only squanders valuable resources but also releases toxic substances into the environment, threatening soil and water quality and, ultimately, human health.

The unsustainable lifecycle of technological products underscores the need for a paradigm shift toward more sustainable practices in technology production, consumption, and recycling. Initiatives to reduce the environmental footprint of technology must focus on several key areas: improving the efficiency and extending the lifespan of products, developing greener sources of energy, and enhancing recycling processes to recover valuable materials. Encouraging the adoption of circular economy principles in the tech industry, where products are designed for durability, repairability, and recyclability, can significantly mitigate the environmental impacts of technology.

Addressing the environmental costs associated with technology is not only a technical challenge but also a moral imperative. As we continue to rely on technological solutions to address global challenges, integrating sustainability into the heart of technological innovation is crucial. By fostering more sustainable practices, we can ensure that our technological advances contribute to a healthier planet, rather than detract from it.

Technology's Role in Climate Change

The relationship between technology and climate change is rich, embodying both the cause of and potential solution to one of the most pressing global challenges of our time. On one hand, technological advancement and industrialization have significantly contributed to climate change through the increased emission of greenhouse gases, resource depletion, and environmental degradation. On the other hand, technology offers promising pathways for mitigating climate impacts and transitioning towards a more sustainable and resilient future.

Renewable energy technologies such as solar, wind, hydro, and geothermal power present viable alternatives to fossil fuels, which are the primary source of carbon emissions driving climate change. By harnessing the earth's natural resources to generate clean energy, these technologies can significantly reduce our carbon footprint and mitigate the effects of global warming. The development and scaling of renewable energy technologies are crucial for achieving the emissions reductions needed to limit global temperature rise to below 1.5°C or 2°C, as outlined in the Paris Agreement.

Smart grids represent another technological innovation with the potential to revolutionize energy consumption and distribution. By integrating digital technology with the electrical grid, smart grids enable more efficient management of energy resources, reducing waste and optimizing the use of renewable sources. This not only enhances the reliability and sustainability of the power supply but also empowers consumers to play an active role in their energy use through real-time information and demand response capabilities.

Carbon capture and storage (CCS) technologies offer a different approach to addressing climate change by capturing carbon dioxide emissions from sources like power plants and industrial processes and storing them underground to prevent their release into the atmosphere. While CCS technologies are still in the developmental and deployment stages, they could play a significant role in decarbonizing sectors where emissions are more challenging to reduce.

Despite the potential of these technologies to contribute to climate mitigation, their deployment must be approached with caution. It is essential to consider the long-term environmental implications and societal impacts of technological solutions to climate change. For instance, the production and disposal of renewable energy systems, the infrastructure for smart grids, and the processes involved in CCS all have environmental footprints that must be carefully managed. Moreover, the transition to a low-carbon economy must be equitable, ensuring access to clean energy technologies across different regions and communities and preventing new forms of environmental injustice.

As we navigate the complex landscape of climate change, technology stands as a double-edged sword. While it has undeniably played a role in exacerbating the problem, it also holds the keys to many solutions. The challenge lies in harnessing this potential responsibly, ensuring that technological innovations are developed and deployed in ways that are sustainable, equitable, and aligned with the long-term health of our planet.

Towards Sustainable Technological Solutions

In the pursuit of progress, the relationship between technology and sustainability has become increasingly complex and interdependent. As the global community faces unprecedented environmental challenges, the imperative for sustainable technological solutions has never been more pronounced. A sustainable approach to technological development is one that harmonizes innovation with environmental stewardship, social equity, and ethical responsibility. It demands a holistic perspective that transcends the immediate allure of technological breakthroughs, focusing instead on the broader implications of these advancements.

The integration of sustainability principles into the very fabric of technological innovation requires a concerted effort across all stages of development. From conceptualization to deployment, each phase should be guided by considerations of how a technology impacts not just the present, but also the future of our planet and its inhabitants. This approach advocates for the minimization of environmental footprints, the promotion of social welfare, and the adherence to ethical standards that respect both human and ecological rights.

Government policies and regulatory frameworks play a pivotal role in fostering a sustainable technological landscape. By implementing regulations that mandate environmental assessments, encourage the use of green technologies, and promote energy efficiency, policymakers can set the stage for responsible innovation. Furthermore, incentives for research and development in sustainable technologies can catalyze breakthroughs that align with environmental goals. These policies, however, must be flexible and adaptable, evolving with the technological landscape to address emerging challenges and opportunities.

Equally important is the cultivation of a societal ethos that values and demands sustainability. Consumers, corporations, and communities alike must champion the cause, advocating for technologies that offer sustainable solutions to our most pressing problems. This collective effort can drive demand for innovations that prioritize long-term wellbeing over short-term gains, encouraging developers and investors to pursue projects with positive environmental and social impacts.

The path towards sustainable technological solutions is fraught with challenges, requiring a delicate balance between innovation, environmental conservation, and social responsibility. Yet, the rewards of this approach are immense, offering a vision of the future where technology serves as a pillar of sustainable development. By embedding sustainability into the heart of technological progress, we can ensure that our advancements today do not come at the expense of tomorrow's world. In doing so, we pave the way for a future where technology and nature coexist in harmony, fostering a healthier, more equitable, and sustainable world for generations to come.

Integrating Technology with Social and Environmental Goals

The task of harmonizing technological advancement with wider social and environmental ambitions calls for a concerted, multidisciplinary effort. This endeavor requires the collective expertise of technologists, policymakers, ethicists, and the public to navigate the intricate challenges that lie at the confluence of technology, society, and the environment. The goal is to cultivate a symbiotic relationship where technological innovation not only propels us forward but does so in a way that directly benefits humanity and safeguards our planet for future generations.

Encouraging a culture of collaboration among these diverse stakeholders is crucial. By breaking down silos and fostering open dialogue, we can uncover synergies between technological capabilities and societal needs. This collaborative approach enables the identification of priorities and the development of technologies that are not just innovative but are also ethical, sustainable, and inclusive.

One of the key strategies in this integration process is the adoption of ethical frameworks and guidelines that govern technological development. These frameworks should emphasize sustainability, equity, and social welfare as core considerations, ensuring that technology contributes positively to societal and environmental objectives. By embedding these values into the DNA of technological innovation, we can steer development towards solutions that address critical global challenges such as climate change, inequality, and health disparities.

Public engagement and education play a vital role in aligning technology with social and environmental goals. Informing and involving the public in discussions about the direction of technological innovation can lead to more democratic and socially responsive technologies. Public awareness and understanding of the potential and limitations of technology can also foster a more critical and informed discourse, leading to better governance and oversight of technological developments.

The application of technology in addressing social and environmental challenges also requires a shift in investment and funding priorities. Supporting research and initiatives that focus on sustainable development, clean energy, and equitable access to technology can drive progress in meeting these goals. This involves not only direct investment in technology development but also in building the infrastructure and capacity needed to deploy these technologies effectively and equitably.

Ultimately, creating an ecosystem where technology serves humanity's best interests involves reimagining the role of technology in society. It means moving beyond the pursuit of innovation for its own sake and towards a model where technological progress is intrinsically linked to sustainable and equitable development. By fostering collaboration across disciplines, adopting ethical frameworks, engaging the public, and redirecting investments towards socially and environmentally beneficial technologies, we can work towards a future where technology truly contributes to the well-being of society and the health of our planet.

The belief in technological salvation as the answer to all societal challenges is an illusion that overlooks the implications of unchecked technological advancement. While technology undoubtedly holds the potential for significant positive impact, its development and application must be guided by rigorous ethical considerations, social equity, and environmental sustainability. By reevaluating our approach to technological progress, we can move towards a future where innovation enhances the human condition without compromising our moral values or the planet's health. This chapter calls for a collective effort to navigate the complexities of the digital age with wisdom, responsibility, and foresight, ensuring that technology truly serves as a catalyst for genuine progress.

Epilogue: A Reflection on Our Era

As we stand at the crossroads of history, looking back at the trail we've traversed and ahead to the paths uncharted, our era presents a paradox of unparalleled progress and unprecedented challenges. The advances in technology, medicine, and science have propelled humanity to heights once unimaginable, crafting a world more connected and informed. Yet, the shadow of these achievements casts long and complex problems—climate change, social inequality, ethical dilemmas born of technological overreach, and a pervasive sense of uncertainty about the future.

This era, marked by digital revolution and global interconnectivity, has redefined human existence, bringing to the forefront questions of identity, privacy, and the very fabric of social cohesion. As we grapple with these challenges, our responses, both individual and collective, will indelibly shape the legacy of our time. The responsibility to wield the double-edged sword of progress with wisdom, to bridge divides rather than deepen them, and to forge a future that honors both the planet and the myriad lives it hosts, rests upon our generation.

The reflection on our era is not merely an academic exercise but a call to action—a reminder that the future is not predetermined but crafted by the choices we make today. It is an invitation to envision a world where progress is measured not by the accumulation of wealth or technological prowess but by the well-being of all its inhabitants and the health of the environment we share.

Afterword: The Historian's Burden

The task of the historian, in chronicling the complexities of our era, is both monumental and fraught with the burden of perspective. To capture the essence of a time so rich in innovation yet riddled with contradictions requires a delicate balance between the objective and the interpretive, between the grand narratives of progress and the undercurrents of resistance and resilience.

Historians of the future will sift through the digital footprints and cultural artifacts of our time, seeking to understand the motivations, aspirations, and challenges that defined us. They will explore how we confronted the ethical quandaries of technological advancement, how we addressed the existential threat of climate change, and how we navigated the shifting landscapes of identity and community in an increasingly virtual world.

In their endeavors, historians carry the burden of contextualizing our achievements and our failings within the broader brush strokes of human history. They bear the responsibility of extracting lessons that can illuminate the path forward, lessons that can guide future generations in facing their own challenges with insight gleaned from the past.

This reflection on our era and the historian's task underscores the profound interconnectedness of all human endeavor. It is a reminder that history is not merely a record of what has been but a mirror reflecting our collective soul, inviting introspection and inspiring a vision of what we might yet become.